Ernst Schering Research Foundation Workshop 14
Apoptosis in Hormone-Dependent Cancers

W0105917

Ernst Schering Research Foundation
Workshop 14

Apoptosis in Hormone-Dependent Cancers

M. Tenniswood, H. Michna
Editors

With 59 Figures and 18 Tables

Springer-Verlag Berlin Heidelberg GmbH

ISBN 978-3-662-03124-7   ISBN 978-3-662-03122-3 (eBook)
DOI 10.1007/978-3-662-03122-3

CIP data applied for

Product liability: The publishers cannot guarantee the accuracy of any information about dosage and application contained in this book. In every individual case the user must check such information by consulting the relevant literature.

Typesetting: Data conversion by Springer-Verlag

21/3135–5 4 3 2 1 0 – Printed on acid-free paper

Preface

To be, or not to be, that is the question:
Whether 'tis nobler in the mind to suffer
The slings and arrows of outrageous fortune
Or to take arms against a sea of troubles
And by opposing end them? To die; to sleep:
No more; and by a sleep to say we end ...

William Shakespeare, Hamlet (III.i)

If we replace the word "opposing" with "apoptosing" in Hamlet's well-known soliloquy and read it in terms of cell biology, it is as if Shakespeare already appreciated the cell suicide process. This type of death, conceptualized as programmed cell death or apoptosis, is a physiological mechanism. Apoptosis contributes to the control of cell turnover and the elimination of cells in normal and (pre)neoplastic tissue both during organogenesis and in completely developed tissue. A complex molecular machinery governs proliferation and most probably a cascade of factors induces programmed cell death or apoptosis. The function of this cell suicide process is probably to delete cells with damaged DNA and its ubiquitous nature has only recently been realized, accounting theoretically for the broad range of application in medicine. Nevertheless, the question of why such a diverse panel of drugs would program cells to actively commit suicide has not been answered. Most interestingly, in seeking to improve the selectivity of cancer therapy, the fact that different cell types possess different thresholds at which they respond to damage with apoptosis seems to offer an attractive option. Thus, hormone-dependent tumor epithelial cells readily engage in apoptosis whereas hormone-independent tumor cells do not. And so, after a century of pharmacology, we are now entering the age of cell physiology in disease treatment.

Fig. 1. The participants of the workshop.
From left to right, first row: K. W. Colston, F. Vignon, N. Kyprianou,
J. Welsh; *second row:* G. J. van Steenbrugge, P. S. Rennie, M. Tenniswood,
G. Aumüller, H. Michna; *third row:* P. R. Walker, K. Parczyk, W. Bursch,
A. Menrad; *fourth row:* D. Henderson, M. Schneider, G. Vollmer

Since the mechanism of apoptosis is unknown, the editors of this monograph thought it would be a challenge to bring experts together to present the latest findings, to foster scientific cooperation, and to coherently organize the known data on the subject of "apoptosis in hormone-dependent tumors."

Whether the reader is knowledgable in the field of apoptosis or is looking for a quick way to catch up on the latest provocative ideas, the editors hope that this volume will provide valuable information.

Martin Tenniswood
Horst Michna

Table of Contents

List of Contributors

K. Akakura
Department of Urology, Chiba University, Chiba, Japan

U. Armato
Department of Histology and Embryology,
Institute of Anatomy and Histology, University of Verona, Verona, Italy

G. Aumüller
Department of Anatomy and Cell Biology, Philipps-Universität,
Robert-Koch-Straße 6, 35033 Marburg, Germany

M. Bacher
Department of Anatomy and Cell Biology, Philipps-Universität,
Robert-Koch-Straße 6, 35033 Marburg, Germany

H. Bonkhoff
Department of Pathology, Universität des Saarlandes, 66424 Homburg-Saar,
Germany

N. Bruchovsky
Department of Cancer Endocrinology, British Columbia Cancer Agency,
University of British Columbia, 600 West 10th Avenue, Vancouver, B.C.,
Canada, V5Z 4E6

W. Bursch
Institut für Tumorbiologie-Krebsforschung der Universität Wien,
Borschkegasse 8a, 1090 Wien, Austria

K. W. Colston
Steroid Biochemistry Group, St. George's Hospital Medical School,
London SW 17 ORE, UK

W. Eicheler
Department of Anatomy, Philipps-Universität, Robert-Koch-Straße 6,
35033 Marburg, Germany

A. Ellinger
Histologisch-Embryologisches Institut der Universität Wien, Ordinariat II,
Schwarzspanierstraße 17, 1090 Wien, Austria

M. Gleave
Department of Cancer Endocrinology, British Columbia Cancer Agency,
University of British Columbia, 600 West 10th Avenue, Vancouver, B.C.,
Canada, V5Z 4E6

S. L. Goldenberg
Department of Cancer Endocrinology, British Columbia Cancer Agency,
University of British Columbia, 600 West 10th Avenue, Vancouver, B.C.,
Canada, V5Z 4E6

B. Grasl-Kraupp
Institut für Tumorbiologie-Krebsforschung der Universität Wien,
Borschkegasse 8a, 1090 Wien, Austria

H. Gourdeau
Apoptosis Research Group, Institute for Biological Sciences,
National Research Council, Bldg M54, Montreal Road, Ottawa, Ontario,
Canada K1A 0R6

R. S. Guenette
W. Alton Jones Cell Science Center, 10 Old Barn Road,
Lake Placid, NY 12946, USA

P. M. Holterhus
Department of Anatomy, Philipps-Universität, Robert-Koch-Straße 6,
35033 Marburg, Germany

S. Y. James
Steroid Biochemistry Group, St. George's Hospital Medical School,
London SW 17 ORE, UK

H. Kienzl
Institut für Tumorbiologie-Krebsforschung der Universität Wien,
Borschkegasse 8a, 1090 Wien, Austria

L. Konrad
Department of Anatomy, Philipps-Universität, Robert-Koch-Straße 6,
35033 Marburg, Germany

T. H. van der Kwast
Department of Pathology, Erasmus University Rotterdam, P.O.Box 1738,
3000 DR Rotterdam, The Netherlands

N. Kyprianou
Division of Urology, Department of Surgery,
University of Maryland Medical Center, 22, South Greene Street,
Baltimore, MD 21201, USA

A. G. Mackay
Steroid Biochemistry Group, St. George's Hospital Medical School,
London SW 17 ORE, UK

H. G. Mannherz
Department of Cell Biology, Philipps-Universität, Robert-Koch-Straße 6,
35033 Marburg, Germany

N. Marceau
Cancer Research Centre, University of Laval, Hôtel Dieu de Quebec, Quebec,
Canada

H. Michna
Research Laboratories, Schering AG, Müllerstraße 170-178, 13342 Berlin,
Germany

M. Mooibroek
W. Alton Jones Cell Science Center, 10 Old Barn Road,
Lake Placid, NY 12946, USA

L. Müllauer
Institut für Tumorbiologie-Krebsforschung der Universität Wien,
Borschkegasse 8a, 1090 Wien, Austria

M. H. A. Oomen
Department of Urology, Division of Urological Oncology,
Erasmus University Rotterdam, P.O.Box 1738, 3000 DR Rotterdam,
The Netherlands

K. Parczyk
Research Laboratories, Schering AG, Müllerstraße 170-178, 13342 Berlin,
Germany

F. Rauch
Department of Cell Biology, Philipps-Universität, Robert-Koch-Straße 6,
35033 Marburg, Germany

H. Renneberg
Department of Anatomy, Philipps-Universität, Robert-Koch-Straße 6,
35033 Marburg, Germany

P. S. Rennie
Department of Cancer Endocrinology, British Columbia Cancer Agency,
University of British Columbia, 600 West 10th Avenue, Vancouver, B.C.,
Canada, V5Z 4E6

C. M. A. de Ridder
Department of Urology, Division of Urological Oncology,
Erasmus University Rotterdam, P.O.Box 1738, 3000 DR Rotterdam,
The Netherlands

H. Rochefort
University of Montpellier I, Department of Cell Biology and INSERM Unit
148 on "Hormones and Cancer", 60, rue de Navacelles, 34090 Montpellier,
France

S. Schenk
Friedrich Miescher Institut, P.O. Box 2543, 4002 Basel, Switzerland

F. H. Schröder
Department of Urology, Division of Urological Oncology,
Erasmus University Rotterdam, P.O.Box 1738, 3000 DR Rotterdam,
The Netherlands

R. Schulte-Hermann
Institut für Tumorbiologie-Krebsforschung der Universität Wien,
Borschkegasse 8a, 1090 Wien, Austria

M. Sikorska
Apoptosis Research Group, Institute for Biological Sciences,
National Research Council, Bldg M54, Montreal Road, Ottawa, Ontario,
Canada K1A 0R6

M. Simboli-Campbell
W. Alton Jones Cell Science Center, 10 Old Barn Road,
Lake Placid, NY 12946, USA

G. J. van Steenbrugge
Department of Urology, Division of Urological Oncology,
Erasmus University Rotterdam, P.O.Box 1738, 3000 DR Rotterdam,
The Netherlands

L. Sullivan
Division of Urology, University of British Columbia, 600 West 10th Avenue,
Vancouver, B.C., Canada, V5Z 4E6

D. Taillefer
W. Alton Jones Cell Science Center, 10 Old Barn Road,
Lake Placid, NY 12946, USA

M. Tenniswood
W. Alton Jones Cell Science Center, 10 Old Barn Road,
Lake Placid, NY 12946, USA

L. Testolin
Department of Histology and Embryology,
Institute of Anatomy and Histology, University of Verona, Verona, Italy

L. Török
Institut für Tumorbiologie-Krebsforschung der Universität Wien,
Borschkegasse 8a, 1090 Wien, Austria

F. Vignon
University of Montpellier I, Department of Cell Biology and INSERM Unit
148 on "Hormones and Cancer", 60, rue de Navacelles, 34090 Montpellier,
France

G. Vollmer
Institut für Biochemische Endokrinologie,
Medizinische Universität zu Lübeck, Ratzeburger Allee 160, 23538 Lübeck,
Germany

P. R. Walker
Apoptosis Research Group, Institute for Biological Sciences,
National Research Council, Bldg M54, Montreal Road, Ottawa, Ontario,
Canada K1A 0R6

W. M. van Weerden
Department of Urology, Division of Urological Oncology,
Erasmus University Rotterdam, P.O.Box 1738, 3000 DR Rotterdam,
The Netherlands

J. Welsh
Department of Biochemistry, University of Ottawa, Faculty of Medicine,
Ottawa, Ontario, Canada

1 Hormonal Control of Prostatic Differentiation and Morphogenesis: The Impact of Apoptosis and Steroid Hormone Receptor Expression

G. Aumüller, P. M. Holterhus, W. Eicheler, H. Renneberg,
M. Bacher, L. Konrad, H. Bonkhoff, F. Rauch,
and H. G. Mannherz

1.1 Introduction

Apoptosis or active cell death (Bursch et al. 1992) is a process whereby cells die in response to specific physiological signals. The morphological sequence of events, as described by Kerr et al. (1972; for review, see Wyllie et al. 1980; Fesus et al. 1991; Fesus 1993) appears to be common to most epithelial cells. They require an expenditure of metabolic energy, active gene expression, and protein biosynthesis. Histologically, the process is characterized by cell shrinking to the extent that they pull away from neighboring cells and the basement membrane, undergoing both nuclear and cytoplasmic condensation. The latter results in the formation of the so-called apoptotic bodies, thought to require increased expression of tissue transglutaminase (Piacentini et al. 1991; Fukuda et al. 1993). Apoptotic bodies are sequestered either by neighboring cells or macrophages, thereby escaping access to the immune system. Chromatin condensation is thought to result from activation of an endogenous Ca^{2+}, Mg^{2+}-dependent endonuclease. Internucleosomal DNA is preferentially digested, resulting in a visual "ladder" of DNA fragments in multimers of 180 base pair units upon electrophoresis, which can be visualized in situ by means of the so-called terminal transferase reaction (Gavrieli et al. 1992).

Cancer cells have lost morphological differentiation to various degrees, but they have regained ancestral stem cell properties, as present in embryonic cells, e.g., a high proliferative capacity and spreading activity due to reduced attachment to basement membranes or loss of polarity. In several embryonic cell types, such features are counterbalanced by programmed cell death, providing embryonic tissues an enormous sculpturing force that gives shape to the developing masses of cells.

In adult tissues, the abnormal onset or blockage of apoptosis has been related to pathological situations. There is now increasing evidence to suggest, for example, that aberrant apoptosis which leads to an imbalance between cell proliferation and cell death may be an important feature in the development of a malignant phenotype in transformed cells. This seems particularly true for the prostate, where apoptosis has been shown to occur as a result of androgen deprivation (Kyprianou and Isaacs 1988). Colombel et al. (1992) have demonstrated that androgen-regulated apoptosis in rat prostate gland results from reentry of differentiated prostate cells onto a defective cell cycle. This process has been identified as a cascade of protooncogene activation (transient induction of c-fos prior to the induction of c-myc transcripts) that usually correlates with cellular proliferative responses in vitro (Buttyan et al. 1989).

In the present study we have analyzed the expression of apoptosis-related cellular signals in the developing human prostate by means of novel histochemical methods to elucidate the potential significance of apoptosis and its hormonal regulation in the prostate during the process of morphogenesis and differentiation. Such knowledge should make it possible to deduce the potential role of these cellular events in malignant transformation of prostate cancer cells.

1.2 Normal Development of the Human Prostate

1.2.1 Fetal Development

Human fetal prostatic differentiation begins with mesenchymal proliferation in the urogenital sinus (Kellokumpu-Lehtinen et al. 1980). Epithelial bud formation occurs in the tenth developmental week. Acinar cells differentiate into "secretory" prostatic cells at the time of highest androgen production in the fetal testis. Kellokumpu-Lehtinen et al. (1980) studied prostatic differentiation in human embryos measuring crown–rump lengths from 43 to 130 mm (corresponding to an age of 9–17 weeks). At the age of 10 weeks, when the verumontanum has developed, histological differentiation begins close to the openings of the mesonephric ducts by outgrowth of several buds of the urethral epithelium into the surrounding mesenchyme. The epithelium of these outgrowths resembles that of the neighboring stratified urethral epithe-

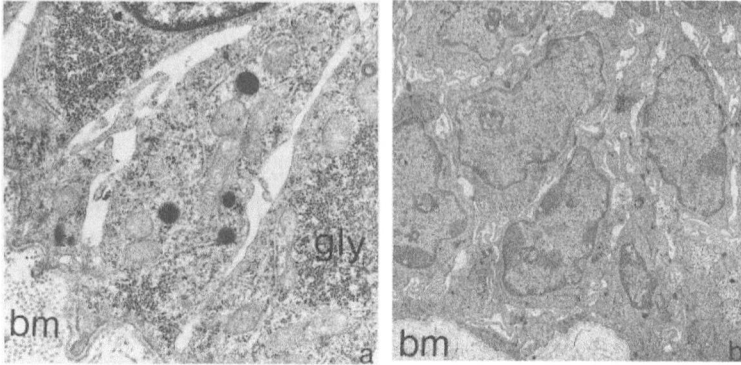

Fig. 1a,b. Ultrastructure of early fetal prostatic epithelium. **a** Prostatic epithelial cells from an embryo of 50 mm crown–rump length (CRL; 11 weeks of gestation, WG). Cells resting on a basement membrane (*bm*) contain much glycogen (*gly*), × 21 000. **b** In an older specimen (125 mm CRL, 17 WG) prostatic anlagen consist of multilayered epithelium forming solid buds, surrounded by a basement membrane (*bm*), × 6000

lium (Fig. 1a). The cells rest on an undulating or folded continuous basement membrane, which separates the epithelium from the surrounding mesenchyme. The initially solid buds acquire a lumen at their terminal or central portions by the end of the 11th week.

Between the 11th and 14th week, when the number of epithelial outgrowths increases, the lumen-containing buds transform into tubuloacinar anlagen. Epithelium of primitive glands consists of layers of three to five cells, most of which are round and apolar. They have numerous slender cytoplasmic processes extending into wide intercellular spaces (Fig. 1b). Some apical cells have become columnar and polarized apicobasally with a large elongate to oval nucleus in the center of the cell; very few contain apical granules. Even though after the 13th week some apical cells become polarized, displaying apical granules, these cells appear to be quiescent when compared to the secretory cells of the mature prostate.

The perinatal increase in squamous epithelial metaplasia, preferentially in the portions of the glandular ducts close to the verumontanum has been related to increased estrogen sensitivity of the prostatic anlagen. Perhaps under the influence of maternal estrogens, during the 15th and 16th weeks, the diameters of the epithelial cords decrease, but otherwise the histological organization and the ultrastructure of the epithelium remain unchanged. Triangular cells, similar to the postnatally observed basal cells, are seen in the basal portion of the epithelium.

1.2.2 Postnatal Differentiation

Whereas Kellokumpu-Lehtinen et al. (1980) found the basal lamina discontinuous in a few places and the epithelium to be in contact with the underlying mesenchymal cells in fetal prostate, we observed a very prominent and thickened basement membrane in postnatal infantile monkey specimens, separating prostatic epithelium from the surrounding stroma. Postnatal development of the human prostate proceeds in three phases: (1) a regression period after birth, (2) a subsequent quiescent period until the onset of puberty at 12–14 years, and (3) a maturation period between 14 and 18 years.

Before and after birth, the collicular portion of the urethral wall contains prostatic gland ducts that display strongly metaplastic stratified squamous epithelium. A few months later, metaplastic epithelium is replaced by a cuboidal pseudostratified epithelium. In the periphery of the developing prostate, epithelial anlagen form solid cords with numerous buds surrounded by a thick basement membrane. Basal cells express basal cell-type cytokeratin and form a kind of pacemaker during morphogenesis of acini. The outgrowth of developing buds starts from these basal cells that are surrounded by a layer of connective tissue and fibrocytes.

In later postnatal development, prostatic epithelial cells gradually acquire the appearance of those of mature prostate. In rat prostate, at the cellular level developmental changes are characterized by a further increase in the amount and complexity of the rough endoplasmic reticulum (RER) and size of the Golgi complex (Flickinger 1971). In the human prostate, functional maturation of the epithelium generally follows the same pattern as in other mammals. In a prostate tissue section from a 14-year-old boy, most glands still have a more or less duct-like appearance. In the prostate of a 15-year-old boy some of the prostatic acini have developed papillary projections which are still rather thick, although some cells appear to be actively secreting. Some of these cells give a strongly positive periodic acid-Schiff (PAS) reaction indicating the presence of glycoproteins. These cells disappear later and are absent in the mature gland. In hyperplastic glands and particularly in certain prostate cancers, they may reappear.

The prostate has achieved its fully mature state when the following criteria are fulfilled:

1. Differentiation of the epithelium into secretory, basal and neuroendocrine cells is finished.
2. PAS-positive mucus cells are lacking in epithelium.
3. All glandular cells are immunoreactive for secretory proteins (see below).

1.2.3 Cellular Compounds

In prostatic development different cell types sequentially appear and disappear, especially in prostatic epithelium (and to a lesser degree in the prostatic stroma). Cells derived from urogenital sinus epithelium, containing characteristic masses of glycogen, transform into undifferentiated squamous epithelial cells. Immediately after lumen formation, adluminal cells develop into polarized columnar cells, whereas undifferentiated "pacemaker" cells reside in the basal layer which eventually form typical triangular basal cells and neuroendocrine cells. These are particularly numerous in the urethral portion of the developing ducts and later in peripheral gland buds, indicating an in situ histogenesis. During perinatal development, periurethral nests of squamous metaplastic epithelium are formed which disappear shortly after birth. With the onset of puberty, the pseudostratified epithelium of the gland ducts is replaced by the terminally differentiated form, i.e., glandular cells, basal cells, and neuroendocrine cells. The shape of the acini reaches a high degree of complexity through developing papillary folds. Changes in the cell populations of prostatic epithelium during development are accompanied by chemical differentiation of the respective cells. Changes in the cellular inventory of the glands are thought to depend on differentiation steps with or without apoptosis of the parental forms.

1.2.4 Chemical Differentiation of the Human Prostate During Development

Kellokumpu-Lehtinen (1983) and Kellokumpu-Lehtinen et al. (1980) demonstrated by enzyme histochemistry the presence of acid phosphatase in urethral and prostatic epithelium throughout the developmental period from the eighth to 14th week, when fetal androgen production begins and reaches its maximum, although estrogen is also present in high amounts (Zondek et al. 1986). During this period, cells with neuroendocrine characteristics already appear in the basal portions of the buds. They display strong chromogranin A, serotonin, and/or calcitonin immunoreactivity. In early developmental stages, acid phosphatase activity is predominantly localized in lysosomes and in the Golgi complex. Later, when some of the prostatic cells become polarized and have an increased amount of RER and a larger Golgi complex, reaction products are also seen in the apical portion of the cells.

By means of specific antibodies against different acid phosphatase isoenzymes (lysosomal vs. secretory), we found that the enzyme histochemical method formerly used visualized the lysosomal (at least the nonsecretory) form of acid phosphatase in these immature glands (unpublished observations). There is no immunoreactivity of secretory acid phosphatase or prostate specific antigen (PSA) present in prenatal prostatic secretory cells. Most of the secretion of the immature glands stains intensely with the PAS reaction und Alcian blue at pH 3.0, indicating the presence of neutral and acidic mucopolysaccharides (proteoglycans). These immature secreting cells are retained until the onset of pubertal maturation of the epithelium. Their fate is still uncertain. The onset of prostatic maturation at the age of 14–15 years is recorded from the appearance of immunoreactivities for secretory acid phosphatase (Aumüller et al. 1983) or PSA. Usually, cells on top of papillary folds or lining the larger ducts in the intermediate portion of the gland start to develop immunoreactivity, which then rapidly spreads all over the glandular system, finally reaching the subcapsular acini which later retain their relatively undifferentiated phenotype until the age of 16–18 years. Whereas secretory cells are rapidly identified by their PSA immunoreactivity, basal cells display strong cytokeratin 5 immunoreactivity, and neuroendocrine cells generally contain chromogranin A (Aumüller 1991). Using these markers, both proliferative activity of the cells and active cell death can be recorded.

1.3 Morphology of Experimentally Induced Apoptosis in the Prostate

Rat ventral prostate was an early object of apoptosis research in hormone-dependent tissues (Kerr and Searle 1973; Sandifort et al. 1984). Sandifort et al. (1984) counted an average of one body per five acini in the prostate of control animals. After castration, the incidence of apoptotic bodies increased markedly during the first 3 days to about seven per acinus and then declined, so that by 13 days they were relatively infrequent.

Using the terminal transferase reaction, epithelial cells containing degrading nuclei are readily identified. In prostatic epithelium, numerous nuclei are labeled 3 days after castration. Fewer labeled nuclei are present in coagulating gland. In control experiments (where no labeled nucleotides are used) no labeling occurs.

Comparing different portions of the prostatic complex on days 1–5 after castration, significant differences in the labeling density are seen (Fig. 2). Rat lateral and ventral prostate react far more intensely than the other lobes do. There are considerable differences between the various lobes of the prostate, not only with regard to the intensity of apoptotic reduction of cell number, but also to the time course of the apoptotic process (Fig. 3). Counting at least 500 nuclei per section, we have quantified the percentage of apoptotic cells in the different lobes of control animals and compared this to the percentage in animals 1, 3, and 5 days after castration. In the ventral prostate of control animals, the number of apoptotic cells was less than 0.1%, on day 1 it was still below 1%, but on day 5 it had reached nearly 8%. In the lateral prostate, the control value of less than 0.1% increased to about 1% on day 1, reached a peak of around 8.5% on day 3 and dropped to around 0. 1% on day 5. A comparable time course was observed in the coagulating gland, but at considerably less intensity, the highest level (on day 3) hardly reaching 2%. In contrast, changes in dorsal prostate and seminal vesicles were minimal and only in the range of 0.1%. The numbers mentioned must be regarded as rough estimates, as only semithin sections of the respective glands from one animal per time point were evaluated. They indicate, however, that the ventral prostate is particularly prone to apoptosis after androgen deprivation. Nevertheless, the question is whether androgen deprivation results in immediate changes in the androgen receptor (AR) mechanism or compensatory changes of the cells.

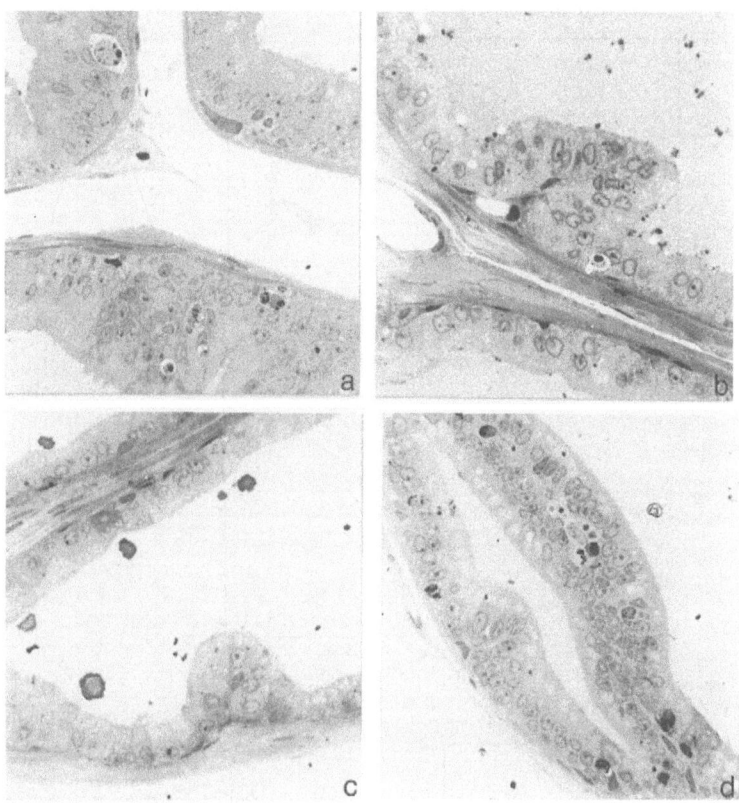

Fig. 2a–d. Semithin sections of different prostate lobes from a rat castrated 3 days prior to sacrifice. **a** Ventral prostate, **b** coagulating gland, **c** dorsal prostate, **d** lateral prostate, × 200

G. Aumüller et al.

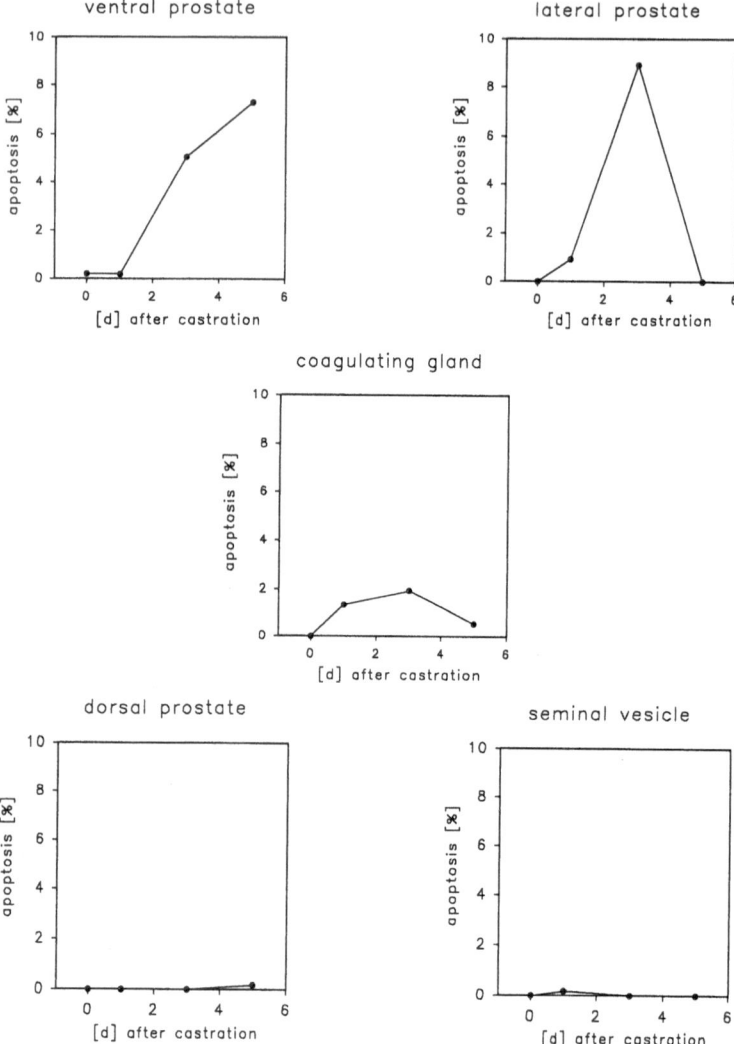

Fig. 3. Semiquantitative evaluation of the time course and intensity of apoptosis in different male accessory sex glands of the rat after castration

1.3.1 Ultrastructure in Castrated Animals

The ultrastructural equivalent of active cell death in ventral prostate of rats 3 days after castration follows the well-established morphological pathway (Wyllie et al. 1980). Individual cells start retracting from the adjacent cells (Fig. 4a), chromatin condenses at the nuclear envelope, and portions of the cytoplasm as well as nuclear fragments are sequestered, forming large membrane bound vacuoles (Fig. 4b) and are di-

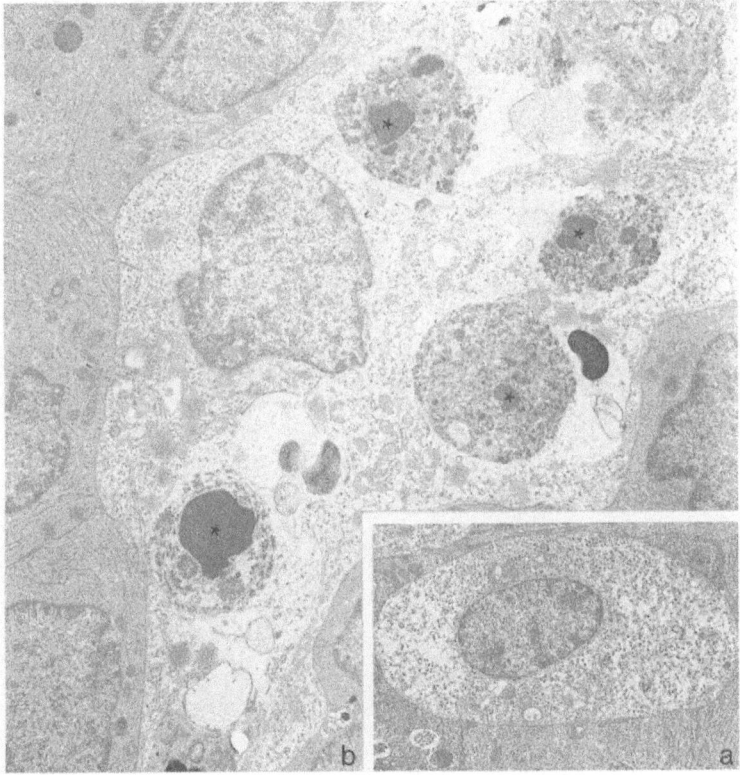

Fig. 4a,b. Ultrastructure of apoptosis in rat ventral prostate 1 (**a**) and 5 (**b**) days after castration. Pycnotic nuclei are indicated by *asterisks*. Magnification in **a** × 2300, **b** × 3000

gested by invading macrophages. Specimens of rat ventral prostate 1, 3, and 5 days after castration therefore served as a positive control in our immunohistochemical studies of human developing prostate specimens.

1.3.2 Differences of Gene Expression in Prostatic Stroma and Epithelium After Androgen Withdrawal

It has recently been demonstrated that the Ca^{2+} Mg^{2+}-dependent endonuclease active during nuclear fragmentation is functionally and antigenically identical to DNase I (Peitsch et al. 1994). We have correlated steroid hormone receptor expression with the expression of DNase and other functional proteins in the prostate of rats at different stages follow-

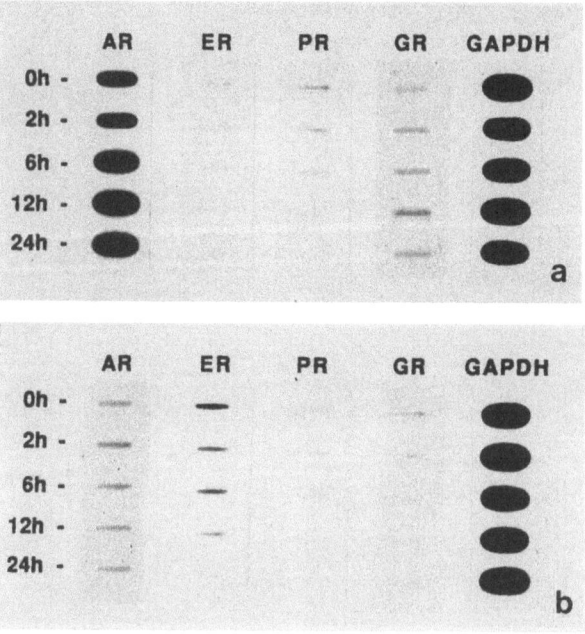

Fig. 5a,b. Slot blot analysis of androgen receptor (*AR*), estrogen receptor (*ER*), progesterone receptor (*PR*), and glucocorticoid receptor (*GR*) mRNA in epithelium (**a**) and stroma (**b**) of ventral prostates of control (0 h) and castrated (2 h–24 h) rats. Calibration with glyceroaldehydephosphate dehydrogenase (*GAPDH*)-cDNA. For technical details, see Bacher et al. (1993)

ing castration, using separate epithelial and stromal fractions (Bacher et al. 1993). In slot blot hybridization analyses of RNA from intact animals, signals for AR and progesterone receptor (PR) were higher in epithelium than in stroma. Glucocorticoid receptor (GR) mRNA concentration in epithelium and stroma were nearly identical, while estrogen receptor (ER) mRNA concentration in stroma was apparently elevated relative to epithelium. After castration, the following changes in steroid hormone receptor mRNA concentration were determined during the first 24 h (Fig. 5): AR mRNA level in epithelium was induced, while it was largely unaltered in stroma. ER mRNA level, predominant in stroma and very low in epithelium, was reduced both in stroma and in epithelium. PR mRNA in stroma was reduced; it was slightly increased in epithelium. Whereas the AR mRNA level in epithelium was induced within the first 24 h after castration, the low stromal signal was slightly elevated in castrated animals after 1 week as demonstrated by northern blot analysis (not shown). In estrogen-treated animals, it was below the control value. No data on AR m RNA in epithelium could be obtained 1 week after castration, due to the low yield of isolated epithelium. The ER mRNA level, predominant in stroma, was clearly reduced within 24 h following castration, while increased signals could be observed in animals which had been treated with estrogen for 1 week. In the stroma of animals which had been castrated 1 week previously, no ER signal was detected at all, as shown by slot blot analysis and polymerase chain reaction (PCR).

To correlate these findings with potential events of apoptosis, we studied the expression of apoptosis-related enzymes in these specimens as well (Fig. 6). Tissue-type transglutaminase, an enzyme thought to be involved in the apoptotic program (Fesus et al. 1987), was constitutively expressed in prostatic stroma. Only minimal changes appeared in the stroma of animals castrated 1 week previously compared to control and estrogenized rats (Fig. 6a). mRNA of DNase I was not detectable in northern blot analyses in intact and estrogen-treated animals, but its expression started 6 h after castration and increased during the following 18 h in epithelium (Fig. 6b). The presumable peak of intensity was obviously not reached within 24 h after castration. These results are not in favor of a significant role of tissue-type transglutaminase in early postcastration involution of the rat prostate, whereas DNase I mRNA apparently is upregulated as early as the early immediate genes (Buttyan et al. 1989).

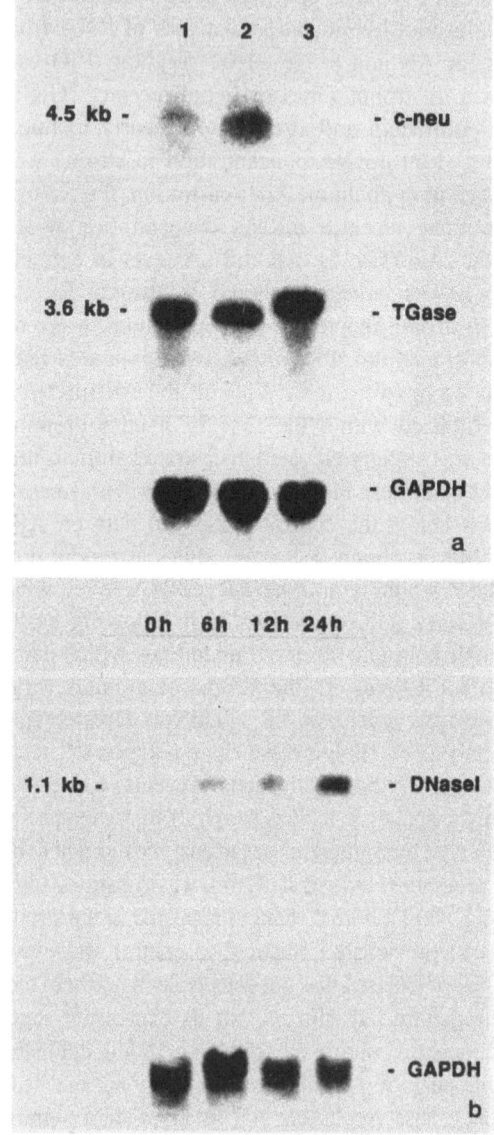

Fig. 6a,b. Legend see p. 15

1.4 Correlation of Apoptosis Markers
with Functional Proteins in Prostate Development

Table 1 gives a survey of the specimens studied, Tables 2 and 3 a synopsis of the relevant results. These are detailed in the following paragraphs.

Table 1. Survey of materials and methods

Fetal prostate

Gestation weeks   19, 20, 26, 27, 29, 30, 34, 36
(Courtesy of Department of Pathology, Freiburg, Germany)

Infantile and pubertal prostate

Months	3 1/2
Years	4, 5–6, 8, 15, 16, 17, 18

(Courtesy of Department of Pathology, Homburg-Saar, Germany)

Adult and pathological prostate

Years	23, 41, 65
BPC	Over 65 untreated; finasteride-treated
PCa	Over 70 untreated; LHRH-agonist, CPA-treated

Immunohistochemical markers

Steroid hormone receptors	Androgen, estrogen
5α-reductase	Isoenzyme I, II
Apoptosis markers	Nuclear morphology, terminal transferase reaction, DNase I, tTgase

BPH, benign prostatic hyperplasia; PCa, prostatic carcinoma; LHRH, luteinizing hormone-releasing hormone; CPA, cyproterone acetate.

◄ **Fig. 6a,b.** Northern blot analysis of tissue type TGase (**a**) in stroma (**a**) and DNase I (**b**) mRNA in epithelium (**b**) of control (**a,** *1*; **b,** *0 h*) and castrated (**a,** 1,2 week; **b,** *6 h, 12 h, 24 h*) and estrogen-treated (**a,** 1 week) rats. *GAPDH,* glyceroaldehydephosphate dehydrogenase. (For technical details, see Bacher et al. 1993)

Table 2. Synopsis of immunohistochemical results in fetal and infantile glands

	AR	ER	5αR-1	5αR-2	tTGase
Fetal glands					
19. WG	–	–	+	(+)	(+)
26. WG	+ met.e.	–	++	(+)	(+)
34. WG	+ met.e.	–	+	(+)	(+)
Infantile glands					
3.5 months	+ met.e.	+ str.	+	(+)	
3 years	–	(+)	+	+	+
5 years	(+) ducts	(+)	+e.++str.	+	(+)
8 years	(+)	(+)	++e.++str.	+	(+)

Abbreviations: met.e., metaplastic epithelium; e., epithelium; str., stroma; –, negative; (+), weak; +, positive; ++, very strong; –/+, partly negative, partly positive; AR, androgen receptor; ER, estrogen receptor; 5αR-1, -2, 5-α reductase types 1 and 2; WG, weeks of gestation.

Table 3. Synopsis of immunohistochemical results in pubertal and adult glands

	AR	ER	5αR-1	5αR-2	tTGase
Pubertal glands					
16 years	+/–	+	+	(+)	–
17 years	–	+	+	+	(+)
18 years	+/–	(+)	+	+	(+)
Adult glands					
23 years	+	(+)	+	+	+/–
BPH	++	+/++	++	+	+/++
BPH treated	–/+	+	(+)	(+)	+
PCa	–/+	–/+	+	–/+	–/+
PCa treated	–/+	–/+	–/+	–/+	–/+

Abbreviations as in Table 2.

1.4.1 Fetal Glands

In fetal glands no ER immunoreactivity was observed and AR immunoreactivity was present only in the nuclei of a few metaplastic epithelial cells. In contrast, the nuclei of nearly all prostatic epithelial cells were stained with an anti-human 5α-reductase type 1 (h5αR-1) isozyme antibody. Human 5α-reductase type 2 isozyme (h5αR-2) immunoreactivity, however, was mostly in the background or control range (Fig. 7). Only few nuclei were labeled with the proliferation marker Ki-67 antigen. Transglutaminase and DNase I were in the background range. No labeling was obtained with the terminal transferase reaction. The luminal portion of all prostatic ducts stained consistently with the prostasome antibody (not shown), even those ducts containing several metaplastic cells.

1.4.2 Infantile Glands

Depending on the fixation quality of the paraffin specimens used, a more or less clear labeling occurred with the monoclonal AR antibody. As in the fetal glands, staining of the metaplastic epithelium was most prominent. In prostate tissue from a 3.5-month-old boy, stromal cells were immunoreactive for ER. In specimens from older children, the epithelium reacted, but clearly less intensely. No labeling was obtained by in situ hybridization using a 714-bp AR cDNA probe (Fig. 7c). The most significant immunoreaction was again achieved with the h5αR-1 antibody (Fig. 7a,f), and somewhat less, with the h5αR-2 (Fig. 7b) antibody. Transglutaminase- and DNase-I-immunoreactive cells were scattered throughout the glands and were somewhat concentrated in the periurethral zone. Terminal transferase reaction (Fig. 7d) and Ki-67 antigen labeling were very varied, labeling densities ranging from 0.1% to 1%. No correlation was found with steroid hormone receptor expression or apoptotic cells.

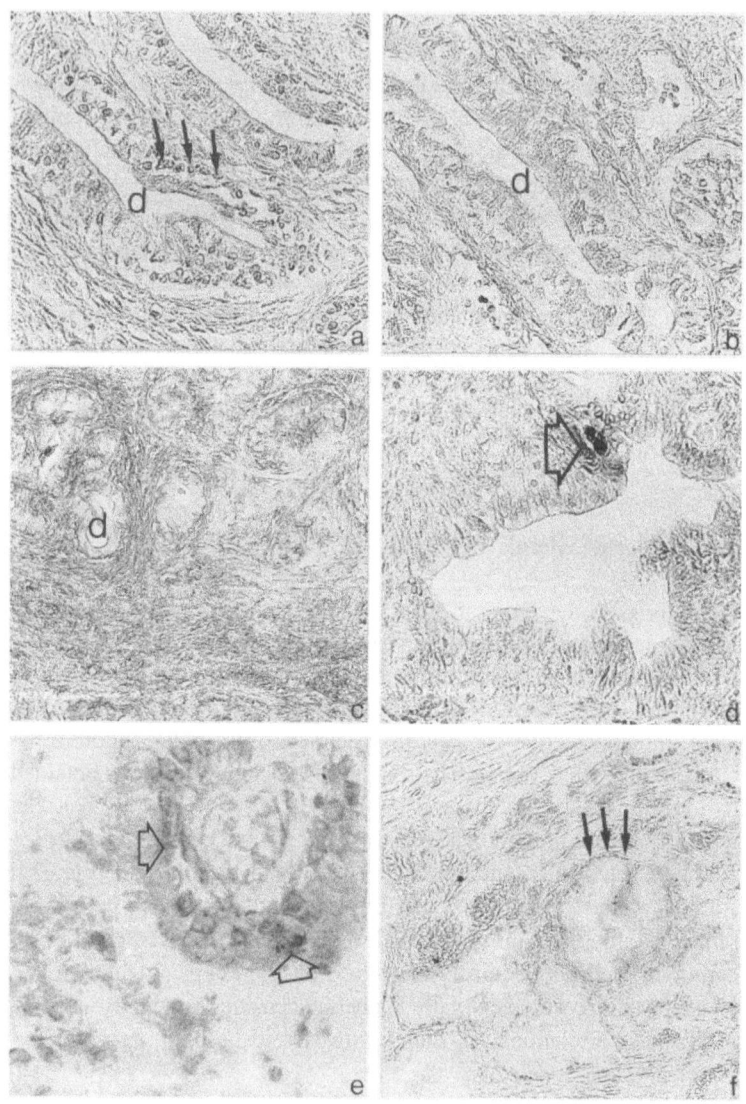

Fig. 7a–f. Legend see p. 19

1.4.3 Pubertal Glands

Tissue preservation was impaired in most of the pubertal gland specimens and, hence, immunoreactivities must be interpreted with great caution. This is particularly true for AR expression, which was mostly at the background level, even if sections had been microwave-treated (5 s, 3 cycles) prior to the immunoreaction. In situ hybridization using a digoxigenin-labeled 714-bp cDNA fragment resulted in a rather homogeneous reaction of the epithelium in the specimens from 17- and 18-year-olds, whereas specimens from younger children were unlabeled (Figs. 8b, 7c). In positively labeled specimens, the staining was concentrated at the base of the epithelium and perinuclear portions of the stromal smooth muscle and vascular endothelial cells. Controls (sense probe controls, RNase treatment, zero control) were consistently negative. Despite the rather homogeneous distribution of the AR and h5αR-1 and h5αR-2 immunoreactivities (Fig. 8a,e), some TGase- and DNase I-immunoreactive areas and cells showing a positive terminal transferase reaction were observed. There was, however, no correlation between these parameters in individual cells. TGase immunoreactivity was very low and was restricted to a few randomly distributed cells. DNase I immunoreactivity was visible in papillary foldings present in the large ducts (Fig. 7e), whereas terminal transferase was frequently positive in urethral epithelium. The latter showed both positive AR and ER immunoreactivity. In the epithelium of the late pubertal prostate (from 17- and 18-year-olds) there were less than 0.1% proliferating and apoptotic cells.

◀ **Fig. 7a–f.** Androgen receptor expression; 5α-reductases and apoptosis in fetal and infantile prostate. **a** Fetal prostate (34th week of gestation), 5α-reductase-1 immunoreactivity (*arrows*) in a ductlike (*d*) acinar anlage. **b** Same specimen; with the 5α-reductase-2 antibody, no immunoreactivity is seen in prostatic duct (*d*) cells. **c** Infantile prostate in a 4-year-old boy; in situ hybridization of the androgen receptor. No specific signal is obtained in prostatic duct (*d*) cells. **d** Infantile prostate fom a 5-year-old boy. One epithelial cell is labeled by the terminal transferase reaction. **e** Late pubertal prostate from a 18-year-old man; DNase I labeling in acinar epithelium covering a papillary fold is indicated by *arrows*. **f** Same specimen shows epithelial nuclei immunoreactive for 5α-reductase-1

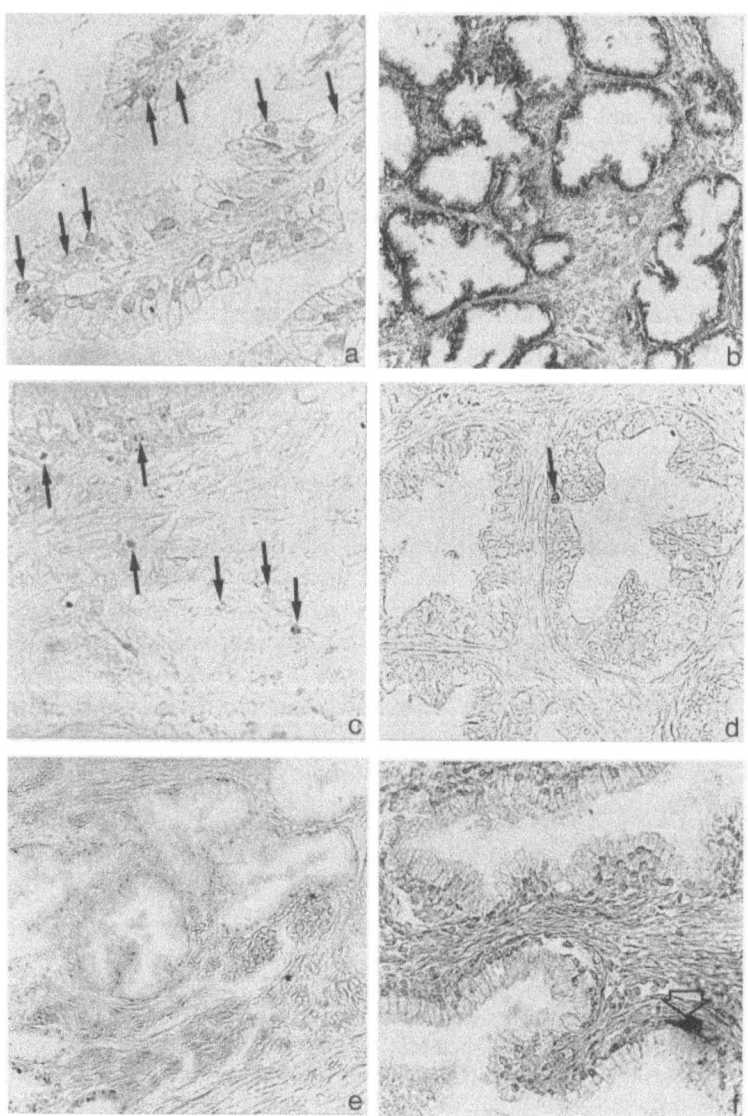

Fig. 8a–f. Legend see p. 21

1.4.4 Adult Prostate, Benign Prostatic Hyperplasia, and Prostatic Carcinoma

In the prostate of normal 23-year-old men, the number of Ki-67 antigen-positive (Fig. 8d) and terminal transferase-labeled cells (Fig. 8f) was extremely low (below 0.01%). Epithelial cells usually contained AR, in periurethral glands also both ER (Fig. 8c) and a strong 5α-reductase immunoreactivity (preferentially isoenzyme 1, Fig. 8e). TGase- and DNase I-immunoreactive cells were rare. This distribution pattern was also present in benign prostatic hyperplasia (BPH) specimens, where immunoreactivities appeared rather strong. Basal cells were more prominent with AR and 5α-reductase immunoreactions. Focal areas with TGase- and DNase-immunoreactive cells were encountered. In prostate cancer (three specimens), divergent results were found in that both strongly positive cells (for either of the antigens studied) and several negative cells were present. The immunoreaction pattern was clearly different from the normal state and will be described elsewhere. The number of terminal transferase-reactive nuclei was elevated compared to the intact glands both in BPH and prostatic carcinoma. As only few specimens have been studied as yet, no deductions are possible from these preliminary observations.

◄ **Fig. 8a–f.** Steroid hormone receptors, 5α-reductase, proliferation, and apoptosis in the adult prostate. **a** Prostatic epithelium from a 23-year-old man shows androgen receptor immunoreactivity in nuclei preferentially of secretory cells; differences in intensity of the immunoreaction are indicated by *arrows*. **b** In situ hybridzation using an androgen receptor cDNA probe (digoxigenin-labeled). A homogeneous reaction of the epithelium is seen. **c** Immunoreactivity of the estrogen receptor in the same specimen. Immunoreactive cells (*arrows*) are mostly present in stroma. **d** Labeling of proliferating cells using the MIB-1 antibody against the Ki-67 antigen (Dianova). One single cell is labeled in acinar epithelium. **e** 5α-reductase-1 immunoreactivity in prostate tissue from an 18-year-old. Colocalization with the androgen receptor (compare with **a**) is evident. **f** Terminal transferase reaction in the young mature prostate. The nature of the reacting cell (*arrow*), presumably a basal cell, is difficult to verify

1.4.5 Survey on Immunoreactivities

1.4.5.1 Markers of Apoptosis and Cell Proliferation

DNase I. DNase I immunoreactivity was present at significant intensity only in late pubertal specimens, where it was localized in the basal portion of cell clusters, forming papillary infoldings in the larger ducts. These areas obviously represent the points where the shaping of the definitive acini takes place. In some specimens a positive terminal transferase reaction was also found there.

Tissue-Type Transglutaminase. Throughout prostatic growth and differentiation only very few cells showed a positive TGase immunoreaction. This was particularly true for metaplastic foci, which were devoid of any TGase immunoreactivity. Only in a few BPH and cancer specimens was the number of TGase-immunoreactive cells increased.

Terminal Transferase Reaction. The terminal transferase reaction was only performed in a few specimens, one fetal and two postnatal. The percentage of reactive nuclei was in the range of 0.1% in the immature glands. In fetal specimens, obviously condensed nuclei indicating apoptosis did not react, perhaps due to partial extraction or excessive denaturation of the tissue.

Proliferation Marker Ki-67 (MIB-1 Monoclonal Antibody). Consistent labeling was achieved only in formaldehyde-fixed fresh cystoprostatectomy specimens (bladder cancer) after extensive microwave treatment (up to 7 cycles) and prolonged antibody incubation. The percentage of labeled epithelial cells ranged between 0.1% and, exceptionally, 1%. Surprisingly, phases of increased growth were not reflected in the labeling density. A few basal and adluminal cells were stained in fetal, infantile, and pubertal glands, but this was very rare. As there are considerable differences in the onset of maturation, the respective proliferative steps may be lacking in our relatively small series of pubertal specimens.

1.4.5.2 Structural and Marker Proteins

Prostasome Antigens. We have prepared an antibody against human prostasomes isolated from ejaculates of healthy voluntary donors which is directed against several prostate antigens in the range of 17–100 kDa (Renneberg et al., in preparation). This antibody stained adluminal plasma membranes of prostatic glandular cells at all stages of development, i.e., immediately after lumen formation in fetal ducts and in metaplastic acini seen in both the perinatal period and in mature secretory cells. Staining of plasma membranes from basal cells was never observed. The staining pattern with the prostasome antibody and the MIB-1 antibody against the Ki-67 antigen supports the idea of a continuous independent renewal of different cell types in the epithelium at relatively slow rates, rather than excessive proliferation and subsequent apoptosis.

Secretory Proteins. Using antibodies against different isoforms of acid phosphatase, we have previously shown that only small amounts of secretory acid phosphatase are synthesized in prepubertal prostates. This finding was confirmed when a commercial antibody against PSA was used. Therefore the prostasome antibody was applied to identify the future secretory cells in prepubertal specimens. Secretion began in the intermediate portions of the larger ducts and then proceeded to the urethral portions and slightly later to the capsular portions. The latter were the last to reach full secretory maturation.

Neuropeptides and Biogenic Amines. The prostate contains a number of endocrine cells producing serotonin, calcitonin, somatostatin, and several other peptides (for review see di Sant'Agnese 1992). Chromogranin A immunoreactivity was used to identify these cells. Obviously, they are already present at very early stages of development, their number being highest in urethral epithelium and then gradually declining to the periphery of the prostatic ducts. In more advanced steps of glandular maturation, the intermediate portion of the ducts contains only few of these cells, whereas in the ductal periphery some branches contain a number of these cells, while others are completely devoid of them. These cells do not demonstrate AR immunoreactivity, although a few express 5α-reductase-1 immunoreactivity and some apparently express the ER. Neuroendocrine cells displaying signs of apoptosis were never observed.

1.4.5.3 Steroid Hormone Effectors

Androgen Receptor. In fetal glands, nuclei with AR immunoreactivity were encountered only in metaplastic areas surrounding the urethra. In infantile and prepubertal specimens AR immunoreactivity was only exceptionally found. In pubertal specimens, however, where AR immunoreactivity was not impressive either, in situ hybridization with an AR probe resulted in a dense and homogeneous reaction.

Estrogen Receptor. ER immunoreactivity was present at low intensity in infantile and moderate intensity in pubertal glands, mostly in the epithelium. In adult glands, localization in the epithelium was rare, but the stroma was stained.

5α-Reductase Isoenzymes. Whereas 5α-reductase-2 immunoreactivity was weak in most specimens (with the exception of the adult gland), 5α-reductase-1 immunoreactivtity, preferentially located within the nuclei of the epithelial cells, was relatively strong throughout all steps of prostatic development. In addition, stromal cells gave a positive immunoreaction in infantile glands. Among all the steroid hormone effectors studied, nuclear 5α-reductase (isoenzyme 1) was the most consistent. Due to its generalized distribution, it was obviously also present in neuroendocrine and in apoptosing cells.

1.5 Discussion and Conclusions

1.5.1 Tissue Quality

All specimens studied were obtained either from autopsies or surgical interventions; as a result they may have suffered a prolonged time of proteolytic degradation. Most were routinely fixed in formaldehyde solution and only a few in Bouin's fluid. The latter, however, turned out to be the less favorable fixative for most antigens and RNA. It must, therefore, be emphasized that the immunohistochemical results presented may reflect a rather strongly impaired functional state. Nevertheless, tissue preservation was sufficiently good to state that apoptosis is a rather rare event in human prostatic epithelium during maturation.

1.5.2 Specificity of the Markers Used

The immunological specificity of the markers prepared by our own laboratories (TGase, DNase I, prostasome antibody, prostatic acid phosphatase isoenzymes, antibodies against h5αR-1,2 isoenzymes) was assessed by western blotting and ELISA systems and was sufficiently high. The commercially available antibodies were used according to the manufacturers' data. Results were rather ambiguous on paraffin sections, although the use of paraffin sections had been indicated by the manufacturers. The question, however, is whether or not marker enzymes such as TGase and DNase I truly reflect certain stages of apoptosis, i.e., which method would be the most suitable for quantification of apoptosis (Sen and d'Incanci 1992). TGase, for example, has been associated with the formation of apoptotic bodies in retinoic acid-treated human cancer cell lines (Piacentini et al. 1991). Retinoic acid per se is a strong inductor of TGase in most cells (Jiang and Kochhar 1992). In human tissues, TGase is preferentially present in cells of mesenchymal origin rather than in epithelia (Thomazy and Fesus 1989). In adult prostate and human prostate cancer cell lines (PC-3, DU-145), no hormonal effects on the expression of tissue-type TGase was observed (Friedrichs et al., in press). Although recommended by different groups as an indicator of apoptotic bodies, we did not find TGase to be useful. The terminal transferase reaction, on the other hand, clearly indicated the occurrence of apoptosis in developing human prostate, although at a very low rate. DNase I immunoreactivity was visible in areas of growing acini. Its distribution pattern was interpreted as indicative of cellular remodeling rather than active cell death. A definite answer, however, cannot be given yet.

1.5.3 Correlation Between Steroid Hormone Effector Expression and Cell Function

On the premise that the immunohistochemical stainings were only partially representative, the assumption is justified that during embryonic development 5α-reductase-1 as well as PR and ER are present in the prostate. AR immunoreactivity was only seen in metaplastic cells. This was not anticipated, as squamous metaplasia can experimentally be induced by estrogen treatment (e.g., in canine prostate). After the onset of puberty, a strong increase in AR content of the epithelium became obvious, indicating the androgen dependence of prostatic maturation. The presence of small amounts of secretory proteins in perinatal prostate (in the absence of immunoreactive AR) is difficult to explain. Either the detection method was insufficient or estrogens are also capable of inducing some basal secretory activity in prostatic epithelium. In long-term castrated and estrogen-treated rats prostatic epithelium also contains cells immunoreactive for prostatic secretory proteins (Zhao et al. 1993). In human pubertal specimens, increased numbers of proliferative and apoptotic cells parallels the appearance of the AR. This was interpreted as a sign of acinar modeling during puberty.

To verify the relationship between AR, 5α-reductase and ER content of the epithelium and its functional state (secretion, proliferation, apoptosis), serial sections were stained with PSA, MIB-1, and DNase as well as the respective receptors. A consistent colocalization was only found for 5α-reductase-1 and the AR, whereas no correlation existed between MIB-1 labeling, and the intensity of PSA and DNase labeling.

Especially in adluminal secretory cells, AR and 5α-reductase stainings were rather weak, but no sign of either apoptosis or reduced secretory activity was observed.

1.5.4 Significance of Apoptosis in the Developing Prostate

In contrast to several other systems, such as the developing limb bud or mammary gland, apoptotic cells are only exceptionally observed in prostatic gland buds during fetal and early postnatal development. During the prepubertal growth phase and particularly during the final development of the gland, proliferative and apoptotic events appear more frequently. Nevertheless, their number is rather low, and only in hyperplastic glands are apoptoses regularly visible, even in simple H&E stained sections.

Another aspect is the regional distribution of actively secreting, proliferating, and apoptotic cells along the prostatic ducts. Rouleau et al. (1990) carefully microdissected the ductal tree of the rat prostate and studied the histological and functional properties of these segments. They found cuboidal nonsecretory adluminal cells covering numerous basal cells in the proximal region, while in the distal region of the ducts active secretion, TRPM-2 expression, and programmed cell death occurs after androgen withdrawal. We have described a similar differentiation gradient in the periurethral portions of the human prostate (Aumüller 1979) and the canine prostate (Aumüller et al. 1980). In contrast to findings from the rat ventral prostate, no such regional differences are present in the distal mature prostate in humans, only in the immature gland. The presence of nonsecretory adluminal cells in peripheral acini is a regular finding in pubertal prostates. No increased apoptoses, however, are found in these immature acini.

1.5.5 Implications for Treatment

Chemotherapeutic agents such as DNA-damaging agents (e.g., melphalan), chemicals interfering with DNA topoisomerase I and II, DNA cross-linkers, agents inhibiting the spindle apparatus, or DNA synthesis have been reported to induce apoptosis in cancer cells (Sen and d'Incalci 1992). On the other hand, castration-induced changes in rat prostate (Kyprianou and Isaacs 1988) have stimulated the idea of induction of apoptosis in hormone-dependent prostate cancer as a means of cancer therapy, based on a kind of cancer cell suicide. The significance of various parameters such as transforming growth factor-β (TGF-β), calcium ions, gene activation, and metabolic energy has carefully been studied (Montpetit et al. 1986; English et al. 1989; Martikainen and Isaacs 1990; Kyprianou and Isaacs 1989; Kyprianou et al. 1989; Hoshikawa et al. 1991; Bettuzzi et al. 1992) in rat prostate. There is, however, no clear-cut proof as yet for the successful application of this concept to human prostate cancer (Bursch et al. 1992). One premise of this concept is that the reduction of the active cell death rate is related to an increased proliferation rate in human prostate cancer. Montironi et al. (1993) have performed a meticulous quantitative study on the occurrence of apoptosis in prostatic intra-epithelial neoplasia (PIN). In low-grade PIN, they found apoptoses at 0.85% in the basal, 0.623% in the intermediate, and 0.474% in the adluminal compartment. Values were clearly higher in high-grade PIN and especially in solid/trabecular adenocarcinomas (2.154% in basal layer and 2.052% in other layers). The authors, therefore, conclude that the previously described increased expression of proliferating cell nuclear antigen (PCNA) is accompanied by increased cell death in prostatic cancer. This observation is well in line with our own studies (on a very small scale as yet), in which we have observed an increased number of nuclei labeled by the terminal transferase reaction and the Ki-67 antigen antiserum in both BPH and prostatic carcinoma. One important aspect with regard to hormonal induction of apoptosis is the presence of androgen-insensitive cells in the prostate, such as the neuroendocrine cells and cells with obviously reduced androgen responsivity (urethral portions of proximal gland ducts). Caution, therefore, is necessary when transposing the findings derived from castrated rats to the situation in human prostate cancer.

1.5.6 Unifying Hypothesis on the Regulation of Prostatic Proliferation and Differentiation

There are a number of observations indicating that the previously developed, simple hypothesis of direct or linear androgen dependence of all the components of the human prostate is too reductionistic a view of a highly complex situation. Instead, there is likely a network of functional interdependencies which has to take into account several structural elements (Fig. 9). These elements interact at different levels: (a) the level of epithelial/stromal interaction, (b) the modulating influence of paracrine factors such as growth factors and/or hormones delivered by various adjacent cell types (which vary along the differentiation gradient of the prostatic ducts), (c) the variable level of supply with androgens as well as other steroids and their metabolites (which change throughout life and depend on the vascular supply of individual acini), and (d) the age-dependent changes in the expression of hormone receptors and activating or degrading enzymes such as 5α-reductases types 1 and 2.

The study of the developing prostate has clearly shown that a sequence of steps occurs within the gland to provide the individual cell with the fully functioning AR mechanism. According to our findings, 5α-reductase-1 is present as a nuclear enzyme already in early fetal development. Later on, 5α-reductase-2 is expressed in the cytoplasm, and still later, the ER is present, whereas the AR is only intermittently expressed – moreover, in cells that are obviously devoid of any se-

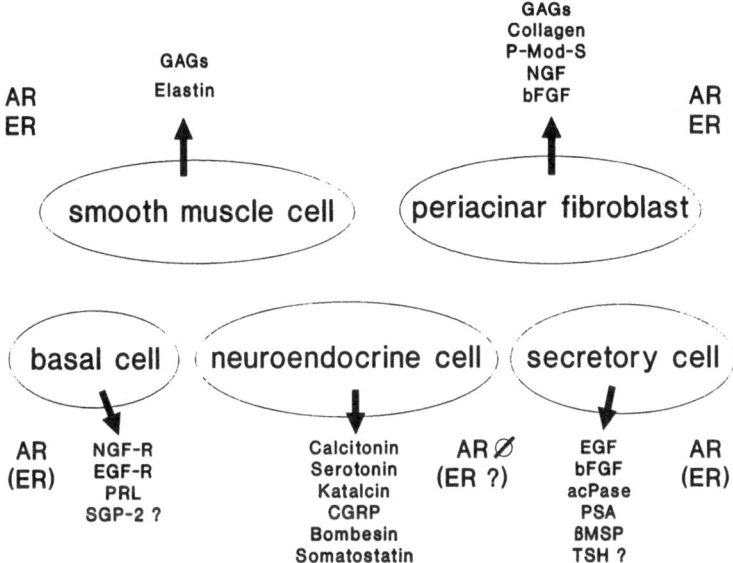

Fig. 9. Cellular components of the human prostate. *Upper panel* shows stromal, *lower panel* epithelial cells. *Large letters* indicate the steroid hormone receptors prevalent in the respective cells; *arrows* point to the most important cellular products. Interaction of the cells can only be sufficiently described in terms of a network with different levels of regulation. *AR*, androgen receptor; *ER*, estrogen receptor; *GAG*, glucosaminoglycans; *P-Mod-S*, peritubular cell factor modifying Sertoli cell function; *NGF*, nerve growth factor; *bFGF*, basic fibroblast growth factor; *EGF*, epidermal growth factor; *PRL*, prolactin; *SGP*, sulfated glycoprotein-2; *CGRP*, calcitonin gene-related protein; *PSA*, prostate specific antigen; β*MSP*, β-microseminoprotein; *TSH*, thyroid-stimulating hormone

cretory differentiation. Only after the AR is present at a high level in acinar cells, is terminal differentiation and functional maturation of the gland achieved. Proliferation is, therefore, not necessarily linked to a complete AR mechanism in the developing prostate.

In addition to cells that invariably enter the state of mandatory androgen dependence, others, such as the neuroendocrine cells, are excluded (Bonkhoff et al. 1993; Krijnen et al. 1993). Whether these cells can transform into androgen-dependent basal or secretory cells

(Bonkhoff et al. 1994a) is still a matter of debate (Bonkhoff et al. 1994b). As there are different levels of androgen sensitivity and responsivity in human prostate cells, it is likely that different levels of regression must also exist, which only develop into active cell death if they pass beyond the point of no return (see Bursch et al. 1992). One may assume that the immunohistochemical demonstration of either tissue-type transglutaminase or DNase I represents such early stages of tissue damage or tissue transformation that are within the normal range of cellular repair mechanisms. Only after excessive cellular changes does the cell reach the point of no return and enter the final steps of apoptosis. Such an excessive level of tissue damage is reached in highly susceptible organs, such as rat ventral prostate, when androgen is withdrawn, whereas in less sensitive organs such as the human gland, a whole series of detrimental events or repeated insults may be required to surrender the prostatic cell to active cell death. In this regard, the functions of the apoptosis-suppressing oncoprotein bcl-2 in human prostate (Colombel et al. 1993) will be of major importance.

Acknowledgements. The financial support of the Deutsche Forschungsgemeinschaft (Grant No: Au-48/10–12) and the Deutsche Krebshilfe (Grant No: W20/93/FU 1; Projekt-Nr. 10316) is gratefully acknowledged.

References

Aumüller G (1979) Prostate gland and seminal vesicles. In: Oksche A, Vollrath L (eds) Handbuch der mikroskopischen Anatomie des Menschen, vol VII/6. Springer, Berlin Heidelberg New York
Aumüller G (1991) Postnatal development of the prostate. Bull Assoc Anat 75 (229):39–42
Aumüller G, Stofft E, Tunn U (1980) Fine structure of the canine prostatic complex. Anat Embryol 160:327–340
Aumüller G, Seitz J, Bischof W (1983) Immunohistochemical study on the initiation of acid phosphatase secretion in the human prostate. J Androl 4:183–191
Bacher M, Rausch U, Goebel HW, Polzar B, Mannherz HG, Aumüller G (1993) II. Stromal and epithelial cells from rat ventral prostate during androgen deprivation and estrogen treatment. Regulation of transcription. Exp Clin Endocrinol 101:78–86

Bettuzzi S, Hiipakka RA, Gilna P, Liao S (1989) Identification of an androgen-repressed mRNA in rat ventral prostate as coding for sulphated glycoprotein 2 by cDNA cloning and sequence analysis. Biochem J 257:293–296

Bonkhoff H, Stein U, Remberger K (1993) Androgen receptor status in endocrine-paracrine cell types of the normal, hyperplastic, and neoplastic human prostate. Virchows Arch [A] 423:291–294

Bonkhoff H, Stein U, Remberger K (1994a) The proliferative function of basal cells in the normal and hyperplastic human prostate. Prostate 24:114–118

Bonkhoff H, Stein U, Remberger K (1994b) Multidirectional differentiation in the normal, hyperplastic, and neoplastic human prostate: simultaneous demonstration of cell-specific epithelial markers. Hum Pathol 25:42–46

Bursch W, Oberhammer F, Schulte-Hermann R (1992) Cell death by apoptosis and its protective role against disease. Trends Pharmacol Sci 13:245–251

Buttyan R, Olsson CA, Pintar J, Chang CS, Bandyk M, Ng P-Y, Sawczuk IS (1989) Induction of the TRPM-2 gene in cells undergoing programmed death. Mol Cell Biol 9:3473–3481

Colombel M, Olsson CA, Ng P-Y, Buttyan R (1992) Hormone-regulated apoptosis results from reentry of differentiated prostate cells onto a defective cell cycle. Cancer Res 52:4313–4319

Colombel M, Symmans F, Gil S, O'Toole KM, Chopin D, Benson M, Olsson CA, Korsmeyer S, Buttyan R (1993) Detection of the apoptosis-suppressing oncoprotein bcl-2 in hormone-refractory human prostate cancers. Am J Pathol 143:390–400

di Sant'Agnese PA (1992) Neuroendocrine differentiation in carcinoma of the prostate. Cancer [Suppl] 70:254–268

English HF, Kyprianou N, Isaacs JT (1989) Relationship between DNA fragmentation and apoptosis in the programmed cell death in the rat prostate following castration. Prostate 15: 233–250

Fesus L (1993) Biochemical events in naturally occurring forms of cell death. FEBS Lett 328:1–5

Fesus L, Thomazy V, Falus A (1987) Induction and activation of tissue transglutaminase during programmed cell death. FEBS Lett 224:104–108

Fesus L, Davies PJA, Piacentini M (1991) Apoptosis: molecular mechanisms in programmed cell death. Eur J Cell Biol 56:170–177

Flickinger CJ (1971) Ultrastructural observations on the postnatal development of the rat prostate. Z Zellforsch 113:157–173

Friedrichs B, Riedmiller H, Goebel HW, Rausch U, Aumüller G (1994) Immunological characterization and activity of transglutaminases in human normal and malignant prostate and prostate cancer cell lines. Urol Res (submitted)

Fukuda K, Kojiro M, Chiu J-F (1993) Induction of apoptosis by transforming growth factor-β1 in the rat hepatoma cell line McA-RH7777: a possible association with tissue transglutaminase expression. Hepatology 18:945–953

Gavrieli Y, Sherman Y, Ben Sasson SA (1992) Identification of programmed cell death in situ via specific labeling of nuclear DNA fragmentation. J Cell Biol 119:493–501

Hoshikawa Y, Satoh Y, Ichii S (1991) Isolation and characterization of cDNA clones for castration-induced mRNAs in the rat ventral prostate. Endocrinol Jpn 38:619–626

Jiang H, Kochhar DM (1992) Induction of tissue transglutaminase and apoptosis by retinoic acid in the limb bud. Teratology 46:333–340

Kellokumpu-Lehtinen P (1983) Localization of acid phosphatase activity in testosterone treated prostatic urethra of human fetuses. Prostate 4: 265–270

Kellokumpu-Lehtinen P, Santti R, Pelliniemi LJ (1980) Correlation of early cytodifferentiation on the human fetal prostate and Leydig cells. Anat Rec 196:263–273

Kerr JFR, Searle J (1973) Deletion of cells by apoptosis during castration-induced involution of the prostate. Virchows Arch [B] 13:87–102

Kerr JFR, Wyllie AH, Currie AR (1972) Apoptosis: a basic biological phenomenon with wide range implications in tissue kinetics. Br J Cancer 26: 239–257

Krijnen JLM, Janssen PJA, Ruizeveld de Winter JA, van Krimpen H, Schröder FH, van der Kwast TH (1993) Do neuroendocrine cells in human prostate cancer express androgen receptor? Histochemistry 100:393–39830

Kyprianou N, Isaacs JT (1988) Activation of programmed cell death in rat ventral prostate after castration. Endocrinology 122:552–562

Kyprianou N, Isaacs JT (1989) Expression of transforming growth factor-β in the rat ventral prostate during castration-induced programmed cell death. Mol Cell Endocrinol 3: 1515–1522

Kyprianou N, English HF, Isaacs JT (1989) Activation of Ca^{2+}-Mg^{2+}-dependent endonuclease as an early event in castration induced prostatic cell death. Prostate 13:103–117

Martikainen P, Isaacs JT (1990) Role of calcium in the programmed death of rat prostatic glandular cells. Prostate 17:175–187

Montironi R, Magi Galluzzi C, Scarpelli M, Giannulis I, Diamanti L (1993) Occurrence of cell death (apoptosis) in prostatic intra-epithelial neoplasia. Virchows Arch [A] 423:351–357

Montpetit ML, Lawless KR, Tenniswood M (1986) Androgen repressed messages in the rat ventral prostate. Prostate 8:25–36

Peitsch MC, Mannherz HG, Tschopp J (1994) The apoptosis endonucleases: cleaning up after cell death? Trends Cell Biol 4:37–41

Piacentini M, Fesus L, Farrace MG, Ghibelli L, Piredda L, Melino G (1991) The expression of "tissue" transglutaminase in two human cancer cell lines is related with the programmed cell death (apoptosis). Eur J Cell Biol 54:246–254

Rouleau M, Léger J, Tenniswood M (1990) Ductal heterogeneity of cytokeratins, gene expression, and cell death in the rat ventral prostate. Mol Endocrinol 4:2003–2013

Sandford NL, Searle JW, Kerr JFR (1984) Successive waves of apoptosis in the rat prostate after repeated withdrawal of testosterone stimulation. Pathology 16:406–410

Sen S, d'Incalci M (1992) Apoptosis: biochemical events and relevance to cancer chemotherapy. FEBS Lett 307:122–127

Thomazy V, Fesus L (1989) Differential expression of tissue transglutaminase in human cells. Cell Tissue Res 255:215–224

Wyllie AH, Kerr JFR, Currie AR (1980) Cell death: the significance of apoptosis. Int Rev Cytol 68:251–306

Zhao GQ, Bacher M, Friedrichs B, Schmidt W, Rausch U, Goebel HW, Tuohimaa P, Aumüller G (1993) Functional properties of isolated stroma and epithelium from rat ventral prostate during androgen deprivation and estrogen treatment. Exp Clin Endocrinol 101:69–77

Zondek T, Mansfield MD, Attree SL, Zondek LH (1986) Hormonal levels in the fetal and neonatal prostate. Acta Endocrinol (Copenh) 112:447–456

2 Apoptosis in Experimental Prostate Cancer

G. J. van Steenbrugge, W. M. van Weerden, M. H. A. Oomen,
C. M. A. de Ridder, T. H. van der Kwast, and F. H. Schröder

2.1 Introduction

2.1.1 Xenograft Models of Human Prostate Cancer

Metastatic prostate cancer is a disease with a high incidence and mortality rate despite advances in early diagnosis and therapeutic intervention (Carter et al. 1990; Schröder 1991). Androgen ablation therapy aiming at reducing tumor burden by inhibition of proliferative activity and inducing programmed cell death (apoptosis) in the tumor tissue is still the current frontline therapy for (advanced) prostate carcinoma (Walsh 1975; Menon and Walsh 1979; Szende et al. 1993). After an initial response, however, tumor relapse occurs due to the growth of androgen-independent prostate cancer cells. This relapse develops even if complete androgen blockade is used, and as a consequence androgen ablation is rarely curative.

The transition of androgen-dependent prostate cancer to an androgen-independent state is a process which is still poorly understood and which can only adequately be studied in experimental model systems which can easily be (hormonally) manipulated (Isaacs and Coffey 1981). Much of the knowledge of the cellular aspects of androgen-regulated growth and progression of prostatic cancer stems from investigations with the normal rat (ventral) prostate and the Dunning R3327 rat prostate cancer system (Isaacs et al. 1978). Although serious doubts were recently raised about the prostatic origin of the Dunning model system (Aumüller et al. 1991), it still has a prominent place in prostate cancer research both in vivo and in vitro. Extrapolation of the results obtained with tumor models in the rat is, however, limited by their nonhuman origin and restricts the direct applicability of such models to the study of human prostate cancer.

Establishment of human prostate cancer cell lines in culture and in vivo, as a heterotransplant in athymic nude mice, is difficult, and generally a very low rate of success has been recognized (Otto et al. 1988). All in all, eight in vitro cell lines, including the androgen-responsive LNCaP model, and seven in vivo xenograft models of human prostate cancer have been described (Van Weerden 1991). In our institution the hormone-dependent PC-82 and two independent tumor lines PC-133 and PC-135 were established more than 10 years ago. In particular the PC-82 has been used for many studies focusing on the role of androgens and estrogens in growth regulation of prostatic carcinoma (Van Steen-brugge et al. 1984; Van Steenbrugge et al. 1988a) and on the regulation of the androgen receptor (Van Steenbrugge et al. 1988b; Ruizeveld de Winter et al. 1992) and proliferation monitoring using the Ki-67 (Gallee et al. 1987). More recent studies with the PC-82 and with another androgen-dependent tumor, named PC-EW, yielded clinically relevant information about the role of low androgen levels and adrenal androgens in the growth of human prostate carcinoma tissue (Van Weerden et al. 1990; Van Weerden et al. 1992).

Recently, our institution was successful in developing a new series of permanent prostate cancer cell lines in vivo. By using athymic nude mice of the NMRI strain instead of the Balb/c strain, a substantial increase in the take rate of human prostate cancer tissues was achieved. Accordingly, during the last 3 years seven new prostatic xenograft models were established, originating from primary tumors (prostatectomy and transurethral resection material) as well as from metastatic lesions (Van Weerden et al. 1994). These tumors represent the various stages of clinical prostate cancer according to differences in the pattern of androgen responsiveness, histological grade of differentiation, expression of the androgen receptor (AR) and of prostatic acid phosphatase (PAP) and prostate specific antigen (PSA).

2.1.2 Apoptosis in the Prostate

Recent studies have shown that tumor growth rate in general is related not only to cell proliferation but also to the rate of (apoptotic) cell death (Dive and Wyllie 1993). Apoptosis or programmed cell death (Wyllie 1980) is a process of major interest and the subject of an increasing number of studies in a variety of normal and malignant tissues including (hormone-responsive) prostate cancer (Isaacs et al. 1992). Androgens, besides having the well-established agonistic ability to stimulate prostate cell proliferation, also have an antagonistic ability to inhibit prostatic cell death (Isaacs 1984). Following castration-induced androgen deprivation the rat ventral prostate rapidly involutes with as many as 80% of cells being lost within the first 10 days after castration (Lee 1981). The death of androgen-dependent ventral prostatic glandular epithelial cells involves a cascade of biochemical changes characteristic of apoptosis (Kyprianou and Isaacs 1988).

Among the various genes that are demonstrated to be implicated in the apoptotic program, also termed the "reactive cascade" (Buttyan et al. 1988), *bcl-2* is known to be an inhibitor of apoptosis (Korsmeyer 1992). There is also growing evidence that the cell cycle suppressor gene *p53* plays an active role in hormonally induced apoptosis in the prostate (Colombel et al. 1992). It is tempting to speculate that the resistance of prostate cancers to undergo apoptosis is a determining factor in the progression of prostate cancer towards androgen independence. The involvement of *bcl-2*, recently described to be highly expressed in the majority of relapsed, i.e., hormone-independent prostate tumors, (McDonnell et al. 1992) is likely. This paper describes the establishment and main characteristics of the newly established human prostate xeno-grafts and it provides information on the occurrence of apoptosis in two of the hormone-dependent models and on the expression of some cell death-associated genes in the panel of models.

In addition, some data are presented concerning androgen-induced cell proliferation and cell death in the hormone-sensitive human pros-tatic tumor cell line LNCaP in vitro. We have previously shown that androgens exert a biphasic response on growth of the LNCaP cell line and that relatively high dosages of androgen inhibited growth of LNCaP cells (Langeler et al. 1993). Phorbol esters have also been shown to induce cell death in the LNCaP cells with features indicative of apop-tosis (Day et al. 1994). In the present study we address the question of whether androgen-induced growth inhibition has the characteristics of apoptosis.

2.2 Material and Methods

2.2.1 Tumor Transplantation

Transplantation of tumor tissue is routinely carried out by implanting small tissue fragments derived from freshly obtained prostate tumor specimens into athymic nude mice (Van Steenbrugge et al. 1984). These concerned primary prostatic carcinomas tissues derived from prostatec-tomy specimens, transurethral resection material from hormone refractory patients, and some metastatic lesions. Tumor tissue was grafted subcutaneously in athymic nude mice of the Balb/c and NMRI strain (derived from the breeding colony of the central animal facilities of Erasmus University). Details about the technique of transplantation, performed under light ether anesthesia, and the methods used to monitor tumor growth have been described previously (Van Steenbrugge et al. 1984), as were the properties of the extensively studied PC-82 and PC-EW tumors (Hoehn et al. 1980; Hoehn et al. 1984; Van Steenbrugge et al. 1988c; Van Weerden et al. 1993). Tumor growth was monitored by the use of caliper measurements of the subcutaneously growing tumor nodules.

2.2.2 Hormonal Manipulation

Hormonally manipulated mice received Silastic implants (Talas, Zwolle, The Netherlands) filled with crystalline steroid, providing constant levels of hormone for extended periods of time (Van Steenbrugge et al. 1984). This method also facilitates hormonal withdrawal and (re)substitution in tumor-bearing castrated male and female mice. Castration was carried out via the scrotal route under anesthesia with tribromoethanol (Aldrich, Beerse, Belgium).

2.2.3 Immunohistochemistry

Tissue sections of formalin-fixed, paraffin-embedded tumor specimens were used for routine histological examination as well as for immunohistochemical staining using an indirect peroxidase–antiperoxidase method with the following monoclonal antibodies directed against PAP, *p53* (antibody DO-7 reactive with wild-type and mutant protein), and *bcl-2* (clone 124), all purchased from Dakopatts, Denmark; F39.4.9, an antihuman AR monoclonal antibody (Ruizeveld de Winter et al. 1991), was provided by the Department of Pathology, Rotterdam, The Netherlands). For staining of *p53*, *bcl-2*, and AR the technique of antigen retrieval using microwave equipment was applied (Shi et al. 1991; Janssen et al. 1994).

For assessment of the human character of each tumor line, tissue sections of tumors from each transplant generation were deparaffinized and stained with bisbenzimide [Hoechst 33258 (4 µg/ml), Sigma, St. Louis, MO, USA]. The differences in fluorescence patterns of Hoechst-binded DNA clearly discriminates between human and murine DNA (Rygaard 1987).

2.2.4 LNCaP Cell Culture

The androgen-sensitive human prostate cell line LNCaP-FGC, which was derived from an early passage of the LNCaP cultures (Horoszewicz et al. 1980), was obtained from Dr. Julius Horoszewicz. The cells were routinely cultured in RPMI-160 medium (Life Technologies, Breda, The Netherlands) supplemented with 7.5% fetal bovine serum (Hyclone, Logan, UT, USA) and glutamine, penicillin, and streptomycin (Van Steenbrugge et al. 1991). Androgen-depleted medium contained 5% dextran-coated charcoal (DCC)-treated serum. For the present experiments LNCaP cells of passage 70–80 were used. The growth characteristics of these (p70) cells under various hormonal conditions have previously been described (Langeler et al. 1993). Androgenic effects on LNCaP growth were tested with the synthetic, nonmetabolizable androgen R1881 (methyltrienolone; New England Nuclear, Boston, MA, USA).

2.2.5 Detection of Apoptosis in LNCaP Cell Cultures

Apoptosis in LNCaP cell cultures was determined by application of the assay of cell viability based on simultaneous staining of LNCaP cell cultures in situ with propidium iodide (PI) and Hoechst 33342 (Polysciences, Warrington, FL, USA). This method provides a means to discriminate between live, necrotic, early, and late apoptotic cells (Pollack and Ciancio 1991). In addition, a recently developed "Cell Death Detection ELISA" (Boehringer, Mannheim, Germany) was applied in an attempt to quantify the occurrence of apoptosis in LNCaP cell cultures. This assay is based on the quantitative in vitro determination of cytoplasmatic histone-associated DNA fragments (mononucleosomes and oligonucleosomes). For both methods, camptothesin (CAM)-treated cells of the human myelogenous leukemic cell line HL-60 were used as controls (Del Bino et al. 1990). Finally, LNCaP cell cultures treated with high dosages of R1881 were processed for transmission microscopy studies.

2.2.6 Miscellaneous

The concentration of PSA in the serum of tumor-bearing mice was measured with an automated enzyme immunoassay (IMX-MEIA, Abbott, IL, USA).

2.3 Results and Discussion

2.3.1 Development of Human Prostate Tumor Xenograft Models

In our institution during a period of more than 10 years (1977–1990) almost 200 clinical specimens were transplanted in Balb/c nude mice, resulting in a very low number of permanent tumor models: the hormone-dependent PC-82 model, two hormone-independent tumors: PC-133 and PC-135 and more recently, the PC-295 tumor (Table 1). The PC-82 was established over 15 years ago and was the first hormone-dependent xenograft model described in the literature and is by far the most extensively studied human xenograft model (Van Steenbrugge 1988; Van Weerden et al. 1990). The other hormone-dependent model, PC-EW, was developed by Hoehn and coworkers (Hoehn et al. 1984) and is also included in the panel of available models presented in Table 1. The very low success rate ($< 5\%$) of prostate carcinomas in nude mice is a generally recognized phenomenon and is thought to be associated with host factors as well as endocrine and/or paracrine (stromal) influences.

About 3 years ago, for technical reasons, we started to use the NMRI strain of nude mice as host animal for heterotransplanting various types of prostatic carcinoma tissues. Remarkably enough, this resulted in a considerably increased take rate: within a period of 2 years six of 19 transplants had a positive take, leading to the development of permanent

Table 1. Main characteristics of a panel of 11 human prostatic carcinoma xenografts

Tumor model	Established	Origin	Androgen dependence	AR	Differentiation[a]	PAP	PSA serum
PC-82	1977	PC	+	+	+	+	+
PC-EW	1981	LN	+	+	±	+	+
PC-133	1981	Bone	−	−	−	−	−
PC-135	1982	PC	−	−	−	−	−
PC-295	1990	LN	+	+	+	++	+
PC-310	1990	PC	+	+	+	++	+
PC-324	1991	TURP	−	−	−	±	−
PC-329	1991	PC	+	+	+	+	+
PC-339	1991	TURP	−[b]	−	−	−	−
PC-346	1991	TURP	+/−	±	−	±	+
PC-374	1992	SM	−	±	±	++	+

PC, primary prostate tumor; LN, lymph node metastasis; TURP, transurethral resection of the prostate; SM, skin metastasis; PAP, prostatic acid phosphatase; PSA, prostate specific antigen.
[a]Glandular differentiation.
[b]No growth in female mice, continued growth after androgen depletion.

tumor lines. The recently (from 1990 to 1992) developed models originate from primary prostatic carcinomas (derived from prostatectomy specimens), from progressive, hormone refractory tumors (tissues derived from transurethral resection material), and from metastatic lesions (lymph nodes as well as skin), as outlined in Table 1. This table contains the origin and main properties of the panel of the total of 11 prostatic xenograft models presently available in our laboratory.

The increased take rate of human prostatic tissues in NMRI nude mice (up to 35%) compared to that in Balb/c mice (less than 5%) seems to be related to host factors. Apparently, such factors are particularly important in the first transplant generation as once the tumors were established as permanent lines they grew equally well in both strains of mice and did not show differences in their characteristics, which will be discussed in the next section.

2.3.2 Properties of the Established Human Prostate Xenografts

The group of seven newly established models was characterized with respect to the differentiation grade, androgen dependence, and expression of the AR and prostate specific markers (PSA and PAP; Table 1). Two subgroups of models can be clearly recognized: differentiated, androgen-dependent tumors expressing the androgen receptor and PAP and secreting PSA in the blood of the host animal (PC-295, PC-310 and PC-295) and two undifferentiated, androgen-independent tumors which are devoid of ARs and do not express or secrete PAP and PSA.

The newly established androgen-dependent tumor models PC-295, PC-310, and PC-329 all show a reduction in tumor volume after androgen ablation. The hormone independent PC-324 and PC-339 models grow equally well in male and female mice. Growth of the PC-374 tumor is not affected by the hormonal status of the host animal, although the tissue of this tumor heterogeneously expressed the AR. The PC-346 tumor is a very interesting model since in some experiments the tumor continues to grow after an initial period of tumor regression following androgen withdrawal, whereas it did not develop when transplanted in female mice. Although not yet definitely established, the PC-346 tumor might be the first human xenograft model showing the clinically so important relapse phenomenon. This independent growth of an initially androgen-dependent tumor clearly will be a very important tool to characterize the origin of and the mechanisms involved in the relapse of originally responsive prostate cancer.

The presented panel of prostate tumor lines provides the opportunity to study and compare the expression of the AR under influence of hormonal deprivation in tissues obtained from successive mouse passages, as was previously done with the PC-82 tumor model (Ruizeveld de Winter et al. 1992). The androgen-dependent AR positive (> 80% of the cells) tumor lines PC-82, PC-EW, PC-295, PC-310, and PC-329 show reduced AR expression (< 30% of the cells) after androgen withdrawal. In the PC-346 and PC-374 tumor lines the AR is heterogeneously expressed (in approx. 30% of the cells), which after androgen ablation is reduced to 10% of AR-positive cells. Even in the androgen-independent tumor lines PC-324 and PC-339 the AR is expressed, although this was restricted to less than 5% of the cells and only when tissue is transplanted in male mice. Interestingly, these tumors and the androgen-independent PC-374 tumor show a further loss of the AR when transplanted in female mice. It has been demonstrated that short-term androgen deprivation of the PC-82 tumor results in downregulation of AR protein but does not affect AR mRNA levels (Ruizeveld de Winter et al. 1992). The influence of long-term androgen depletion on AR mRNA and protein expression has not yet been studied in the PC-82 or any of the other available models. Such data could provide insight into whether (persistent) AR expression (either mutated or not) is associated with the development of androgen independence.

2.3.3 Cell Death in the PC-82 and PC-EW Tumor Models

The volume of PC-82 and PC-EW tumors declines after castration of tumor-bearing mice, the regression of the PC-EW tumor being faster than that of the PC-82 tumor (half-life of 6 and 18 days, respectively (Fig. 1). The resultant decline of the PC-82 tumor nodules after castration is associated with a decrease in cells incorporating BrdU (Van Weerden et al. 1993). Concomitantly, an increase in the number of apoptotic bodies is observed, which reaches its maximum at 4 days post-castration (Table 2). These observations are in agreement with

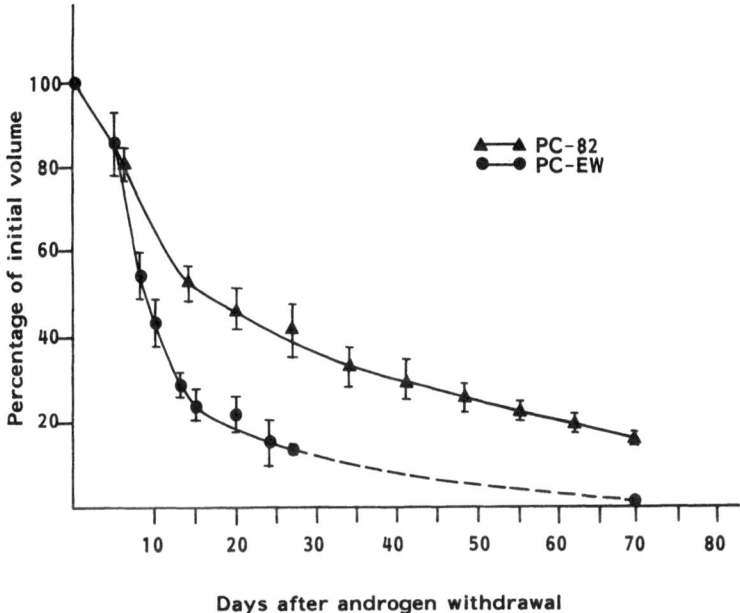

Fig. 1. Patterns of regression after androgen withdrawal from PC-82 ($n = 8$) and PC-EW ($n = 7$) tumor-bearing nude mice

previously reported results of the PC-82 tumor (Kyprianou et al. 1990). The regressing PC-82 tumors do not show any sign of necrotic death, whereas androgen depletion of the androgen-dependent PC-EW tumor induces both apoptosis and necrosis. Regressing PC-EW tumors show widespread necrosis with clusters of apoptotic cells (Table 2). As a consequence of the necrotic cell death the PC-EW tumors regress completely (within 3–4 weeks), leaving small nodules consisting of fibrotic tissue. Regressing PC-EW tumors cannot be restimulated by androgens even after a relatively short period (2 weeks) of androgen depletion. By contrast, in regressing PC-82 tumors viable cells remain, which can be restimulated to grow by androgens even after long-term (over 6 months) androgen depletion (Van Weerden et al. 1993). Spontaneous, that is, androgen-independent, regrowth of these long-term, androgen-deprived PC-82 tumors has never been observed, however. This androgen-sensitive but "apoptosis-resistant" PC-82 tumor cell population possibly represents an intermediate step between androgen dependence and independence and, therefore, may provide a good model to investigate the transition of androgen-dependent cancer cells to endocrine resistance.

2.3.4 Expression of Some Cell Death-Related Gene Products in Prostate Xenografts

Preliminary results obtained on the expression of some functional markers in nine of our xenograft models are shown in Table 3. In order to demonstrate *p53* expression, the monoclonal antibody DO-7 directed against the wild-type and mutant molecule was used. In the PC-324 and PC-339 tumors overexpression of *p53* was noted, that is, nuclei of over 90% of the cells were intensely stained. In the other tumors scattered *p53* immunopositive cells were found, which was assumed to reflect the expression of wild-type *p53* associated with proliferative activity. Interestingly enough, the same two hormone-independent tumors with *p53* overexpression were also strongly immunoreactive for *bcl-2*. Although the *bcl-2* expression may be a differentiation-related phenomenon, in these tumors with *p53* overexpression the intense *bcl-2* staining is most likely the result of a constitutive *bcl-2* expression. It is tempting to speculate about the regulatory influence of *p53* on *bcl-2* as it was very recently demonstrated to occur in other systems (Miyashita et al. 1994). Except for the PC-324 and 339 tumors, the slow growing, highly differentiated, and hormone-dependent PC-310 tumor in the majority of cells also show (faint) immunoreactivity with the *bcl-2* monoclonal antibody, which disappears after androgen withdrawal. Double staining

Table 2. Cell death in regressing PC-82 and PC-EW tumors

Time after castration (days)	PC-82 % apoptosis	PC-82 % necrosis	PC-EW % apoptosis	PC-EW % necrosis
0	1.3 ± 0.6 (5)	< 10	1.6 ± 0.9 (5)	10
4	10.8 ± 3.0 (5)	< 10	5.4 ± 2.6 (5)	30
7	7.2 ± 3.0 (4)	< 10	2.9 ± 1.1 (4)	30
15	2.8 ± 0.2 (3)	< 10	2.4 ± 1.0 (5)	50
30	1.9 ± 1.0 (5)	< 10	2.2 ± 1.3 (4)	80

Data are expressed as mean ± SD, with number of samples in parentheses.

Table 3. Functional markers involved in cell proliferation and cell death of human prostate xenografts

Tumor model	Androgen dependence	Hormonal status	p53	bcl-2
PC-82	+	M	–	–
		–M		–
PC-135	–	F	–	–
PC-295	+	M	–	–
		–M	–	–
PC-310	+	M	–	+
		–M		–
PC-324	–	M	+	+
		F	+	+
PC-329	+	M	–	–
PC-339	–	M	+	±
		F	+	±[a]
		F	+	±
PC-346	+/–	M	–	–
		–M		–
PC-374	–	M	–	–
		F		–

F, female; M, male; –M, castrated male.
[a]Heterogeneous pattern of staining.

of the AR and *bcl-2* reveals that there was no coexpression of the two markers in these tumors (result not shown). Apoptotic cells, which are infrequently seen in sections of almost all tumor tissues, are negative for the *bcl-2* protein.

2.3.5 Androgen Induces Cell Death in LNCaP Cell Cultures

In steroid-depleted (DCC) medium, FGC cells continued to grow at a decreased rate. Maximal growth was induced with the addition of 0.1 nM of the synthetic, nonmetabolizable androgen R1881, whereas the addition of 10 nM R1881 to DCC medium leads to a decrease in cell number (Langeler et al. 1993). In these high-dose, R1881-treated cultures a high rate of cell death is found. It appears that all dying cells are

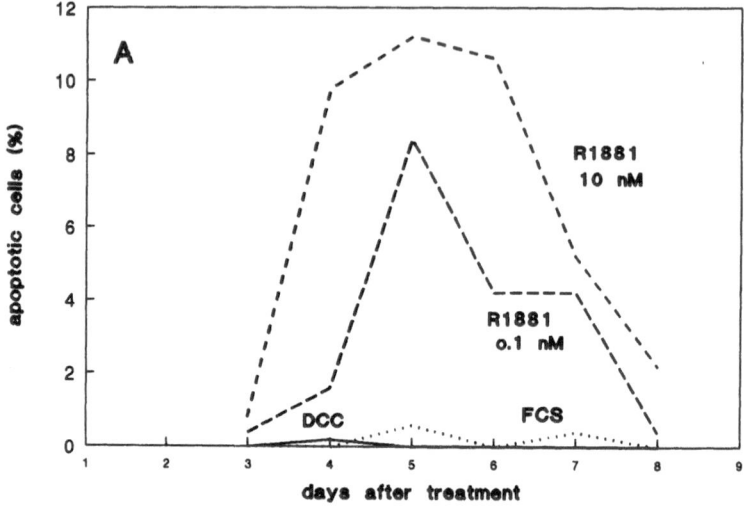

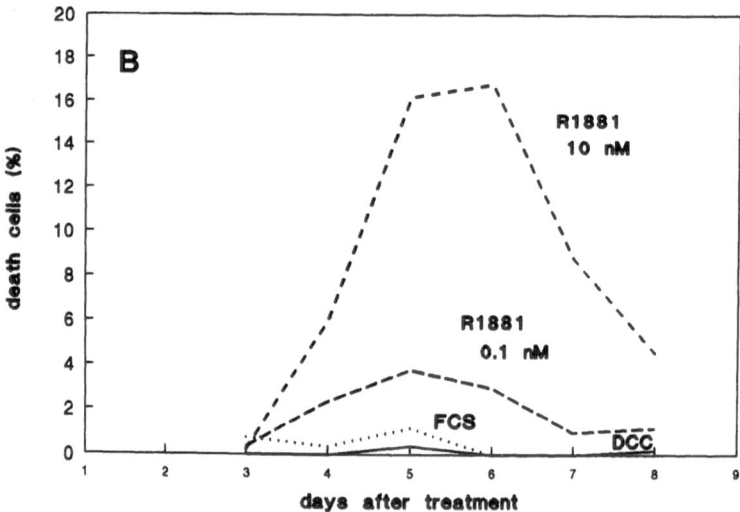

Fig. 2a,b. Apoptotic cell death in LNCaP cells cultured in medium with complete serum (*FCS*), androgen-depleted serum (*DCC*), and DCC supplemented with 0.1 and 10 nM of the synthetic androgen R1881. Cultures were stained with propidium iodide (**a**) and the DNA binding dye, Hoechst 33324 (**b**), respectively detached and that attached cells are vital and continue to grow. Staining of such cultures in situ with a combination of the DNA binding dye, Hoechst 33324, and PI clearly demonstrates that the detached cells have features indicative for apoptosis. Specifically, cytoplasmic contraction, condensation of nuclear chromatin, and the formation of membrane-bound "apoptotic bodies" are observed. Apoptotic events are also confirmed by electron microscopic examination of high-dose, R1881-treated cells. Interestingly, in the R1881-treated cultures initially blue, i.e., intact, PI excluding apoptotic cells were found, whereas gradually cultures also showed increasing amounts of cells with red luminescence that have lost their membrane integrity, but which also had apoptotic features. Apparently, with this method early and late apoptotic cells can be detected.

Estimates were made of the percentage of LNCaP cells that were recognized as being apoptotic based upon their pattern of DNA Hoechst stain and the number of dead cells that had a positive stain with PI. From the time-course experiment shown in Fig. 2 it can be seen that androgen-induced apoptosis is a relatively slow process. The peak of 18% apoptotic cells in 10 nM R1881-treated cells is reached about 6 days after start of treatment (Fig. 2a). From this graph it can be seen that in cultures grown in 0.1 nM R1881, a dosage which optimally stimulates growth of LNCaP cells, a relatively high percentage of apoptotic cells was also found. By contrast, cultures grown in medium with either complete serum or androgen-depleted (DCC) serum have a very low number of apoptotic cells. This appears to be spontaneous cell death occurring in all cell cultures. Withdrawal of androgens from the rat ventral prostate or from the androgen-dependent human PC-82 tumor in vivo results in the induction of apoptosis (Kyprianou and Isaacs 1988; Kyprianou et al. 1990). This apparently does not apply to the androgen-sensitive human LNCaP cell line in vitro, as in androgen-depleted medium growth of the cells was only retarded.

The patterns of occurrence of PI-positive cells under the different hormonal conditions (Fig. 2b) parallels those found for apoptotic cells with yet intact membranes (Fig. 2a), except that the level of dead cells found in cultures treated with 0.1 nM R1881 is considerably lower than that observed in cells cultured in the high concentration of 10 nM R1881.

Table 4. Androgen (R1881)-induced cell death in the LNCaP prostatic cell line

Conc. R1881	enrichment factor[a] (DCC=0)		
	Day 5	Day 6	Day 7
100 nM	12.9	37.0	37.7
10 nM	2.7	59.7	73.1
1 nM	1.3	29.7	21.9
0.1 nM	2.9	43.0	19.3

[a]Enrichment factor of nucleosomes in the cytoplasma of LNCaP cell treated with different concentrations of the synthetic androgen R1881.

The occurrence of apoptosis in the LNCaP cell cultures can also be quantified by the use of a photometric enzyme immunoassay that has recently become available. This test is based upon the determination of cytoplasmic histone-associated DNA fragments present in cells that undergo the process of apoptosis. Samples of the time-course experiment discussed above with LNCaP cultures were measured in this cell death detection ELISA. The nucleosome enrichment factor, relative to that found for the DCC cultures cells, clearly increases in time and in a dose-dependent manner (Table 4). Although increased nucleosome concentrations, i.e., DNA fragmentation, were also found in the cultures with low androgen concentrations, again, with this method the highest rates of apoptotic cell death are observed with 10 nM R1881.

The observed phenomenon of R1881-induced apoptosis might be mediated by the AR, as it has previously been shown that in LNCaP cells treated with high doses of R1881, the AR was downregulated (Langeler et al. 1993). On the other hand, it has been demonstrated in this study that the entire culture of high-dose, R1881-treated cells does not die, and a surviving fraction of cells continued to growth after 3 weeks of exposure to R1881. Whether this represents a heterogeneity of LNCaP cells with different sensitivities for R1881 still needs to be confirmed. Preliminary experiments with high dosages of the natural androgen dihydrotestosterone (DHT) did not result in increased levels of apoptotic cells. Although such cultures received a daily dosage of 100 nM DHT, it must be taken into account that LNCaP cells have been demonstrated to have a high metabolism of both testosterone and DHT.

As a consequence, it cannot be excluded that the phenomenon of androgen-induced apoptosis is an exclusive effect of the synthetic androgen R1881. In the LNCaP cell line, irrespective of the concentration, R1881 induced cell proliferation as well as cell death. The phenomenon could clearly be demonstrated with the DNA staining dye Hoechst-33324 in combination with PI and further quantified with the cell death detection ELISA.

2.4 Conclusions

The recently developed series of human prostatic tumor xenografts together with the previously generated tumor models represents all aspects of clinical prostate cancer: primary hormone-dependent carcinoma, progressive (relapsed) hormone-independent prostate cancer, and metastatic disease. This unique panel of prostatic tumors may result in a better understanding of the biochemical pathway involved in active cell death and of the mechanisms implicated in progression of clinical prostate cancer, i.e., the transition from androgen-dependent to androgen-resistant prostate cancer.

Androgen-induced cell death in the LNCaP prostatic carcinoma cell line was demonstrated to be apoptotic cell death. This as yet incompletely understood phenomenon seems to be restricted to the synthetic androgen R1881.

References

Aumüller G, Gröschel-Stewart U, Altmannsberger M, Mannherz HG, Steinhoff M (1991) Basal cells of H-Dunning tumor are myoepithelial cells. Biochemistry 95:341–349

Buttyan R, Zakeri Z, Lockshin R, Wolgemuth D (1988) Cascade induction of c-fos, c-myc, and heat shock 70 K transcripts. Mol Endocrinol 2:650–657

Carter BS, Carter HB, Isaacs JT (1990) Epidemiologic evidence regarding predisposing factors to prostate cancer. Prostate 16:187–197

Colombel M, Olsson CA, Ng PY, Buttyan R (1992) Hormone-regulated apoptosis results from reentry of differentiated prostate cells onto a defective cell cycle. Cancer Res 52:4313–4319

Day ML, Zhao X, Wu S, Swanson PE, Humphrey PA (1994) Phorbol ester-induced apoptosis is accompanied by NGFI-A and *c-fos* activation in androgen-sensitive prostate cancer cells. Cell Growth Differ 5:735–741

Del Bino G, Skierski JS, Darzynkiewicz Z (1990) Diverse effects of camptothecin, and inhibitor of topoisomerase I on the cell cycle of lymphocytic (L1210, MOLT-4) and myelogeneous (HL-60, KG1) leukemia cells. Cancer Res 50:5746–5750

Dive C, Wyllie AH (1993) Apotosis and cancer chemotherapy. In: Hickman and Tritton (eds) Frontiers in pharmacology and therapeutics. Blackwell, Oxford, pp 21–56

Gallee MPW, Van Steenbrugge GJ, Ten Kate FJW, Schröder FH, Van der Kwast TH (1987) Determination of the proliferative fraction of a transplantable hormone dependent, human prostatic carcinoma (PC-82) by monoclonal antibody Ki-67: potential application for hormone therapy monitoring. J Natl Cancer Inst 79:1333–1340

Hoehn W, Schroeder FH, Riemann JF, Joebsis AC, Hermanek P (1980) Human prostatic adenocarcinoma: some characteristics of a serially transplantable line in nude mice (PC-82). Prostate 1:95–104

Hoehn W, Wagner M, Riemann JF, Hermanek P, Williams E, Walther R, Schrueffer R (1984) Prostatic adenocarcinoma PC-EW, a new human tumor line transplantable in nude mice. Prostate 5:445–452

Horoszewicz JS, Leong SS, Kawinsky E, Karr JP, Rosenthal H, Ming Chu T, Mirand EA, Murphy GP (1983) LNCaP model of human prostatic carcinoma. Cancer Res 43:1809–1818

Isaacs JT (1984) Antagonistic effect of androgen on prostatic cell death. Prostate 5:545–557

Isaacs JT, Coffey DS (1981) Adaptation versus selection as the mechanism responsible for the relapse of prostatic cancer to androgen ablation therapy as studied in the Dunning R-3327-H adenocarcinoma. Cancer Res 41:5070–5075

Isaacs JT, Heston WDW, Weismann RM, Coffey DS (1978) Animal models of the hormone-sensitive and -insensitive prostatic adenocarcinomas, Dunning R3327H, R3327HI and R3327AT. Cancer Res 38:4353–4359

Isaacs JT, Lundmo PI, Berges R, Martikainen P, Kyprianou N, English HF (1992) Androgen regulation of programmed death of normal and malignant prostatic cells. J Androl 13:457–464

Janssen PJ, Brinkmann AO, Boersma WJ, Van der Kwast (1994) Immunohistochemical detection of the androgen receptor with monoclonal antibody F39.4 in routinely processed, paraffin-embedded human tissues after microwave pre-treatment. J Histochem Cytochem 42:1169–1175

Korsmeyer SJ (1992) Bcl-2: an antidote to programmed cell death. Cancer Surveys 15:105–118

Kyprianou N, Isaacs JT (1988) Activation of programmed cell death in the rat ventral prostate after castration. Endocrinology 122:552–562

Kyprianou N, English HF, Isaacs JT (1990) Programmed cell death during regression of PC-82 human prostate cancer following androgen ablation. Cancer Res 50:3748–3753

Langeler EG, Van Uffelen CJC, Blankenstein MA, Van Steenbrugge GJ, Mulder E (1993) Effect of culture conditions on androgen sensitivity of the human prostatic cancer cell line LNCaP. Prostate 23:213–223

Lee C. (1981). Physiology of castration-induced regression in rat prostate. In: Murphy GP, Sandberg AA, Karr JP (eds) The prostate cell: structure and function, Part A. Liss, New York, pp 145–159

McDonnell TJ, Troncoso P, Brisbay SM, Logothetis C, Chung LWK, Hsieh JT, Tu SM, Campbell ML (1992) Expression of the protooncogene Bcl-2 in the prostate and its associations with emergence of androgen-independent prostate cancer. Cancer Res 52:6940–6944

Menon M, Walsh PC (1979) Hormonal therapy for prostatic cancer. In: Murphy GP (ed) Prostatic cancer. PSG Publishing, Littleton, pp 175–200

Miyashita T, Krajewski S, Krajewski M, Gang Wang H, Lin HK, Liebermann DA, Hoffman B, Reed JC (1994) Tumor suppressor p53 is a regulator of bcl-2 and bax gene expression in vitro and in vivo. Oncogene 9:1799–1805

Otto U, Wagner B, Klöppel G, Baisch H, Klosterhalfen H (1988) Animal models for prostate cancer. In: Klosterhalfen H (ed) Endocrine management of prostatic cancer. Walter de Gruyter, Berlin, pp 29–37 (New developments in biosciences, vol 4)

Pollack A, Ciancio G (1991) Cell cycle phase-specific analysis of cell viability using Hoechst 33342 and propidium iodide after ethanol preservation. In: Darzynkiewicz Z, Crissman HA (eds) Flow cytometry. Academic, San Diego, CA, pp 19–24

Ruizeveld de Winter JA, Trapman J, Vermey M, Mulder E, Zegers ND, van der Kwast TH (1991) Androgen receptor expression in human tissues: an immunohistochemical study. J Histochem Cytochem 39:927–936

Ruizeveld de Winter JA, Van Weerden WM, Faber PW, Van Steenbrugge GJ, Trapman J, Brinkmann AO, Van der Kwast TH (1992) Regulation of androgen receptor expression in the human heterotransplantable prostate carcinoma PC-82. Endocrinology 131:3045–3050

Rygaard K (1987) A rapid method for identification of murine cells in human malignant tumours grown in nude mice. In: Rygaard J, Brunner, Graen N, Spang-Thomsen M (eds) Immune-deficient animals in biomedical research. Karger, Basel, pp 268–272

Schröder FH (1991) Endocrine therapy: where do we stand and where are we going? Cancer Surveys 11:177–194

Shi SR, Key ME, Kalra KL (1991) Antigen retrieval in formalin-fixed, paraffin-embedded tissues: an enhancement method for immunohistochemical staining based on microwave oven heating of tissue sections. J Histochem Cytochem 39:741–748

Szende B, Romics I (1993) Apoptosis in prostate cancer after hormonal treatment. Lancet 342:1422

Van der Kwast TH, Schalken J, Ruizeveld de Winter JA, Van Vroonhoven CCJ, Mulder E, Boersma W, Trapman J (1991) Androgen receptors in endocrine-therapy-resistant human prostate cancer. Int J Cancer 48:189–193

Van Steenbrugge GJ, Groen M, Romijn JC, Schröder FH (1984a) Biological effects of hormonal treatment regimens on a transplantable human prostatic tumor line (PC-82). J Urol 131:812–817

Van Steenbrugge GJ, Groen M, de Jong FH, Schroeder FH (1984b) The use of steroid-containing silastic implants in male nude mice: plasma hormone levels and the effect of implantation on the weights of the ventral prostate and seminal vesicles. Prostate 5:639–647

Van Steenbrugge GJ, Groen M, Van Kreuningen A, De Jong FH, Gallee MWP, Schroeder FH (1988a) Transplantable human prostatic carcinoma (PC-82) in athymic nude mice: III. Effects of estrogens on the growth of the tumor. Prostate 12:157–171

Van Steenbrugge GJ, Bolt-de Vries J, Blankenstein MA, Brinkmann AO, Schroeder FH (1988b) Transplantable human prostatic carcinoma (PC-82) in athymic nude mice: II. Tumor growth and androgen receptors. Prostate 12:145–156

Van Steenbrugge (1988c) Transplantable human prostate cancer (PC-82) in athymic nude mice: a model for the study of androgen-regulated tumor growth. Doctoral Thesis, University of Rotterdam

Van Steenbrugge GJ, Van Uffelen CJC, Bolt J, Schröder FH (1991) The human prostatic cancer cell line LNCaP and its derived sublines: an in vitro model for the study of androgen sensitivity. J Steroid Biochem Mol Biol 40:207–214

Van Weerden WM, Van Steenbrugge GJ, Van Kreuningen A, Moerings EPCM, De Jong FH, Schröder FH (1990) Effects of low testosterone levels and of adrenal androgens on growth of prostate tumor models in nude mice. J Steroid Biochem Mol Biol 37:903–907

Van Weerden WM (1991) Animal models in the study of progression of prostate cancer and breast cancer to endocrine independency. In: Berns PMJJ, Romijn JC, Schröder FH (eds) Mechanisms of progression to hormone-independent growth of breast and prostatic cancer. Parthenon, Carnforth, pp 55–70

Van Weerden WM, Van Kreuningen A, Elissen NMJ, De Jong FH, Van Steenbrugge GJ, Schröder FH (1992) Effects of adrenal androgens on the transplantable human prostate tumor PC-82. Endocrinology 131:2909–2913

Van Weerden WM, Van Kreuningen A, Elissen NMJ, Vermey M, De Jong FH, Van Steenbrugge GJ, Schröder FH (1993) Castration-induced changes in morphology, androgen levels and proliferative activity of human prostate cancer tissue grown in athymic nude mice. Prostate 23:149–164

Van Weerden WM, De Ridder CMA, Romijn JC, Van Steenbrugge GJ, Van der Kwast TH, Schröder FH (1994) Characterization of seven newly established human prostate tumor models in NMRI nude mice. Proc Am Assoc Cancer Res 35:282

Walsh PC (1975) Physiological basis for hormonal therapy in carcinoma of the prostate. Urol Clin North Am 2:125–140

Wyllie AH, Kerr JFR, Currie AR (1990) Cell death: the significance of apoptosis. Int Rev Cytol 86:251–306

3 Therapeutic Significance of Apoptosis in the Treatment of Androgen-Dependent and Androgen-Independent Prostate Cancer

N. Kyprianou

3.1 Introduction

3.1.1 Prostate Cancer: The Magnitude of the Problem

Prostate cancer is the most commonly diagnosed malignancy in men in the United States and a major contributor to cancer mortality. The disease has a remarkably high annual mortality rate, with 38 000 men dying each year from metastatic prostate cancer (Boring et al. 1993). Due to the rapid rise in aging of the US population, it is estimated that by the year 2000 there will be a 37% increase in prostate cancer deaths (Carter and Coffey 1990). Approximately 200 000 new cases of prostate cancer are clinically diagnosed each year. The disease varies widely in its clinical aggressiveness. In some patients, prostate cancer metastasizes rapidly, killing the patient within a year of initial clinical presentation, whereas other patients may live for many years with localized disease without apparent metastases. At the same time, prostate cancer is often characterized as being clinically "silent" in a large number of men who harbor it histologically. An inviting challenge in prostate cancer research, therefore, is to distinguish prostatic tumors destined to progress to lethal metastatic disease from those with little likelihood of causing morbidity.

Metastatic prostate cancer is still a fatal disease for which no therapy is available that effectively increases survival. Androgen ablation has been the standard form of therapy for advanced prostate cancer for over 50 years. Despite an initial transient response to androgen ablation therapy, however, the majority of patients relapse to a state unresponsive to further antiandrogen treatment (Carter and Coffey 1990). Androgen ablation monotherapy is rarely curative because it is only targeted against the androgen-dependent prostate cancer cell populations within individual tumors, leaving the androgen-independent clones totally unaffected. After the initial positive response, therefore, due to the death of the androgen-dependent cells, the surviving tumor progresses to an androgen-unresponsive state, rendering the patient's disease no longer curable. To affect all the heterogeneous prostatic cancer cell populations within an individual patient effective chemotherapy specifically targeted against the preexisting androgen-independent cancer cells must be combined with androgen ablation (Raghavan 1988).

The growth of any tumor is determined by the relationship between its rate of cell proliferation and cell death. Successful treatment for androgen-independent cancer cells can be obtained by lowering the rate of cell proliferation and/or by raising the rate of cell death to a point where this exceeds the rate of cell proliferation.

3.1.2 Apoptosis: The Molecular Mechanism of Cell Death

The death of a cell can occur via one of two stereotypic processes, which are biochemical and morphologically distinct (Wyllie et al. 1980). Necrosis is elicited by any of a large variety of factors that lead to the permeabilization of the plasma membrane, with the resultant osmotic lysis of the cell and its internal membranes. In necrotic cell death, the cell does not actively participate in the process of death, but rather it is killed by a hostile microenvironment. In contrast, in programmed cell death or apoptosis, the cell actively participates in the initiation of its own death in response to specific signals in an otherwise normal microenvironment. Apoptosis, classified as the physiologically relevant mode of cell death, is an active, energy-dependent process involving a temporally distinct series of biochemical and molecular steps (Wyllie et al. 1980). Once initiated, programmed cell death leads to a cascade of biochemical and morphological events that result in the irreversible fragmentation of the genomic DNA and then of the cell itself (Wyllie et al. 1980; Umansky et al. 1981; Cohen and Duke 1984). In programmed cell death, fragmentation of DNA into nucleosomal oligomers irreversibly commits the cell to die and occurs prior to changes in plasma and internal membrane permeability. This DNA fragmentation has been shown to result from activation of a Ca^{2+} Mg^{2+}-dependent endonuclease present with the cell nucleus which selectively hydrolyzes DNA at sites located between nucleosomal units, thus resulting in a stereotypic nucleosomal ladder of DNA fragments (Wyllie et al. 1980; Umansky et al. 1981; Cohen and Duke 1984; Kyprianou et al. 1988). This nuclease activation is thought to be triggered by a sustained elevation in the intracellular free Ca^{2+} concentration initiated early in the apoptotic process (Nicotera et al. 1986; McConkey et al. 1989). In necrotic death, the DNA is degraded into a continuous spectrum of sizes as a result of the simultaneous action of both lysosomal proteases and nucleases released in dead cells (Kyprianou and Isaacs 1989a). The morphological pathway for programmed cell death is stereotypic and has been given the name "apoptosis" to distinguish this actively initiated process (i.e., early nuclear changes followed by eventual nuclear disintegration and fragmentation of the dying cell into a cluster of membrane-bound apoptotic bodies) from the passively initiated process of necrotic cell death in which plasma, mitochondria, and lysosomal membrane changes precede disintegration of the cell nucleus (Wyllie et al. 1980). This process of selective cellular deletion occurs widely throughout nature in diverse organized tissue reactions such as embryonic development, morphogenesis, regulation of organ homeostasis in adult life, hormone-induced tissue involution, and cell-mediated immunity (Raff 1992). The clinical significance of apoptosis has only recently been appreciated in efforts to generate more effective therapeutic modalities for cancer treatment (Kyprianou et al. 1991a).

3.2 Apoptosis as a New Strategy for Prostate Cancer Therapy

The potential therapeutic implications of simultaneous activation of apoptosis in androgen-dependent and androgen-independent prostatic cells are clearly very important in the development of cancer treatment modalities for advanced prostate cancer. Such modalities should include effective chemotherapeutic means targeted against androgen-independent cells, which in combination with androgen ablation will activate the apoptotic cascade among all the heterogeneous tumor cell populations, ultimately increasing patient survival.

3.2.1 Apoptotic Response of the Normal Prostate to Androgen Ablation

Androgens have the dual ability to stimulate cell proliferation and inhibit cell death of normal prostatic glandular epithelial cells (Isaacs 1984; Kyprianou and Isaacs 1988). Androgen ablation induces a series of discrete biochemical events that lead to a cessation of cell proliferation and the activation of programmed cell death (apoptosis) of these androgen-dependent prostatic cells, ultimately resulting in the involution of the gland (Kyprianou and Isaacs 1988; Kyprianou et al. 1988). Within 12 h after castration of adult male rats, serum testosterone concentration falls to below 2% of the intact control value. This rapid decline in serum levels of androgen comprises the apoptotic signal to which the androgen-dependent prostate epithelial cells respond by activating their genetic program of cell death. This is an active energy-dependent process which involves a cascade of temporally distinct biochemical and morphological events sequentially leading to fragmentation of DNA into nucleosomal oligomers, induction of expression of TRPM-2/clusterin gene (Léger et al. 1987), transforming growth factor (TGF)-β_1 gene (Kyprianou and Isaacs 1989b), c-*fos*, and c-*myc* protooncogenes (Buttyan et al. 1988), and eventual histological appearance of apoptotic cells in the involuting prostate epithelium (English et al. 1989).

3.2.2 Apoptosis of Human Prostate Cancer Cells Following Androgen Ablation

In vivo studies have identified apoptosis as the molecular mechanism underlying the regression of androgen-dependent human prostatic tumors after androgen ablation (Kyprianou et al. 1990), i.e., androgen-responsive human prostate cancer cells, PC-82, like normal rat prostatic epithelial cells (Kyprianou and Isaacs 1988) are able to undergo programmed cell death in response to androgen ablation. The PC-82 human prostate adenocarcinoma is highly androgen responsive when grown as a xenograft in nude mice (Van Steenbrugge et al. 1984). If intact male nude mice are inoculated with human PC-82 prostatic cancer, continuously growing tumors are produced. Castration of mice bearing PC-82 xenografts (approx. 0.5 cm^3 in size), leads to a significant decrease in the rate of cell proliferation (sevenfold) and an increase in the rate of cell death (11-fold). As a result, the tumor rapidly involutes, reaching approximately half its starting size within 3 weeks following androgen ablation. Biochemical analysis during this regression of the PC-82 tumor demonstrated that both TGF-β_1 and TRPM-2 mRNA levels as well as tumor DNA fragmentation into nucleosomal-sized pieces are detectably increased within the first day following castration (Kyprianou et al. 1990). The levels of all these parameters increase to a maximum on the third day following castration. Histological appearance of apoptotic bodies is a characteristic early event that precedes the dramatic reduction in tumor volume following androgen ablation. If exogenous androgen is given back to the castrated host, DNA fragmentation ceases, TGF-β_1 and TRPM-2 mRNA transcript levels drop, regression stops, and tumor growth resumes.

3.2.3 Apoptosis of Androgen-Independent Human Prostate Cancer Cells Following Treatment with Ionizing Irradiation

Radiation therapy has been shown to be effective in organ-confined disease (Paulson et al. 1982). The use of radiation in advanced prostate cancer is, however, limited to palliation of bone pain and acute neurologic decompensation associated with spinal cord compression (Zietman et al. 1994). Radiotherapy for the treatment of prostate cancer has long been known as an alternative medical therapeutic approach, but the molecular mechanism involved in radiation-induced toxicity in prostatic tumors is poorly defined.

The antitumor effect of ionizing irradiation on human prostate cancer cells was investigated in a series of in vitro studies, using two androgen-independent prostate cancer cell lines, DU-145 and PC-3, as model systems. Exposure of both cell lines to increasing doses of ionizing irradiation (200, 400, 600, 800, 1000 cGy) resulted in a significant induction of cell death, in a dose-dependent manner (Sklar et al. 1993). Temporal analysis of the molecular events involved in this radiation-induced toxicity revealed the characteristic fragmentation of DNA into a nucleosomal ladder (a hallmark of apoptosis) and enhanced expression of specific apoptosis-associated genes (TRPM-2 and TGF-β_1), with both events preceding the dramatic decrease in cell number. These results clearly suggest that radiation-induced cell death proceeds via activation of the apoptotic pathway. Radiation-induced apoptosis has been demonstrated in several other systems, including rat thymocytes (Umansky et al. 1981) and murine tumors (Yamada and Ohyama 1988; Stephens et al. 1991). Evidence emerging from studies on other tumor cell lines support the concept that upregulated release of TGF-β_1 could assist in programming cancer cells for the active suicide process (Yanagihara and Tsumuraya 1992). The temporal induction of the TGF-β_1 gene during radiation-induced cell death points to a role of this negative growth factor as a potential molecular regulator of radiation-induced apoptosis of androgen-independent prostate cancer cells.

3.2.4 Effect of Suramin on Radiation-Induced Apoptosis of Prostate Cancer Cells

Therapeutic modalities in the treatment of prostate cancer directed at interference with growth factors or growth factor receptors are scarce. Suramin, a polysulfonated napthylurea, was originally used for the treatment of parasitic disorders (Tanon and Janot 1924). Suramin, which is a potent inhibitor of viral reverse transcriptase (De Clerg 1979), has recently surfaced as a potential antineoplastic agent on the basis of its ability to exert a cytostatic effect on a variety of solid tumors, including prostate carcinoma (Berns et al. 1990; Stein et al. 1989; Peehl et al. 1991). Suramin has been shown to exert its antineoplastic activity by blocking the binding of certain growth factors to their membrane receptors (Kim et al. 1991), thus inhibiting the mitogenic activity of such growth factors. Other evidence implicates interruption of cellular respiration and changes in intracellular Ca^{2+} as potential mechanisms involved in the antitumor activity of suramin (Rago et al. 1991; Seewald et al. 1991). The use of suramin in the treatment of advanced prostate cancer has been limited by its significant side effects and a relatively narrow therapeutic window. Moreover, its administration is complicated by significant patient variability of measured pharmacokinetic parameters. Revised dosing regimens have been reported and recent studies point to a survival advantage in patients with metastatic prostate cancer treated with suramin (Myers et al. 1992; Eisenberger et al. 1993).

In vitro studies have demonstrated that treatment of androgen-independent prostate cancer cells with suramin over a 5-day period resulted in a dose-dependent sustained inhibition of cellular proliferation for both cell lines examined, i.e., PC-3 and DU-145 (Sklar et al. 1993). These results confirm previous observations demonstrating that suramin causes a dose- and exposure-related cytostatic effect on prostate cancer cell growth (Kim et al. 1991).

Pretreatment of cells with suramin (300 µg/ml) for 3 days prior to exposure to ionizing irradiation had a significant inhibitory effect on radiation-induced apoptosis in both prostate cancer cell lines, while the cytostatic action of the drug predominated, i.e., suramin partially abrogated the cytotoxic effect of irradiation (Sklar et al. 1993). Further kinetic studies demonstrated that suramin treatment of cells (300 µg/ml) 24 h following exposure to ionizing irradiation (1000 cGy) resulted in a time-related dramatic enhancement of radiation-induced cell death (approximately 40%). By 5 days after irradiation with continuous exposure to suramin, more than 70% cytotoxicity was observed for both prostate cancer cell lines. Suramin pretreatment of cells aborted the radiation-induced DNA fragmentation, a molecular effect that correlates with the substantial suppression by suramin of radiation-induced cell death. These observations imply a role for suramin in rendering human prostate cancer cells relatively radioresistant. A possible underlying mechanism for this effect could be proposed on the basis of the hypothesis that suramin exerts its effect in the G_0/G_1 phase of the cell cycle (Kim et al. 1991), while radiation-induced damage occurs in the G_2 phase of the cell cycle (Painter 1980). If indeed suramin causes G_0/G_1 arrest, this would prevent cells from progressing to the G_2 phase, therefore becoming less sensitive to the effect of radiation. Alternatively considering the evidence that elevation of c-H-*ras* oncogene expression has been associated with increased radioresistance in tumor cells (McKenna et al. 1990; Hermens and Bentvelen 1992), it is tempting to suggest that suramin-induced radioresistance of human prostate cancer cells could be mediated by induction of *ras* oncogene expression.

In marked contrast, suramin treatment subsequent to ionizing irradiation substantially enhanced the kinetics of DNA fragmentation as well as of TGF-β_1 and TRPM-2 gene induction, i.e., incubation with the drug after irradiation accelerated the apoptotic process. The enhancement of radiation-induced apoptosis by suramin treatment after irradiation could be explained by the fact that maximal radiation effect is not observed immediately after irradiation, but rather occurs several cell divisions later (Painter 1980). Alternatively, the combined antitumor effect of suramin plus irradiation may be due to a potential suramin interference with the capacity to repair radiation-induced damage, a hypothesis that would support the role of suramin in inhibiting DNA polymerase activity (Jindal et al. 1990). In dual consideration, these results suggest that while preirradiation suramin treatment of prostatic tumors may not confer any therapeutic advantage, postirradiation suramin treatment significantly potentiates the decrease in cell number induced by either treatment alone. The relative timing of this combination treatment is critically important in determining their individual effects on cell-cycle regulation. Prostate cancer patients previously treated with suramin would have minimal response to subsequent radiotherapy, while radiotherapy followed by suramin treatment may yield enhanced effectiveness.

3.2.5 Apoptosis in Androgen-Independent Human Prostate Cancer Cells Undergoing Thymineless Death

Additional in vitro studies have demonstrated that androgen-independent human prostatic cancer cells are able to undergo programmed cell death in response to the nonandrogen-ablative drugs 5-fluorodeoxyuridine (5-FdUr) and trifluorothymidine (Kyprianou et al. 1994). These fluorinated pyrimidines exert their primary cytotoxic effect through indirect inhibition of thymidylate synthetase, leading to depletion of intracellular thymidine-5-triphosphate pools and ultimately resulting in a toxic "thymineless" state (Grem 1990). Simultaneous exposure of prostate cancer cells to a sufficient level of exogenous thymidine completely abrogated the cytotoxicity induced by the fluorinated pyrimidines ($100 \, \mu M$), directly supporting the concept that this cytotoxic effect is a result of the generation of "thyminelesss" state (Kyprianou et al. 1994). The molecular events characteristic of apoptosis, i.e., DNA fragmentation into the nucleosomal ladder and enhanced expression of TRPM-2/SGP-2 (sulfated glycoprotein-2) and TGF-β_1 mRNA transcripts after exposure to either drug, preceded the loss of cell viability, a phenomenon pointing to the apoptotic nature of this fluoropyrimidine-mediated toxicity. Elevated TRPM-2/SGP-2 transcripts have been specifically associated with induction of apoptosis in a number of systems undergoing programmed cell death in response to a wide spectrum of stimuli (other than hormonal ablation), establishing this gene as a useful molecular marker of apoptosis (Léger et al. 1987; Kyprianou and Isaacs 1989b; Kyprianou et al. 1991b; Sklar et al. 1993).

As the search for a functional role of this protein in the initiation of the apoptotic mechanism continues, it is tempting to speculate a potential role of the TRPM-2 gene in the regulation of cell membrane permeability. Such a role would imply its dynamic involvement in Ca^{2+} influx, required for $Ca^{2+}Mg^{2+}$-dependent endonuclease activation (Martikainen et al. 1991). Direct support for a such a concept emerges from the identification of the TRPM-2 encoded protein as the sulfated glycoprotein-2 (Bettuzi et al. 1989), which is involved in an acrosomal reaction during sperm penetration of the egg – a process initiated by an influx of extracellular Ca^{2+} (Sylvester et al. 1984). In view of the membrane localization of the SGP-2 protein (Sensibar et al. 1992), one could speculate on a central involvement of TRPM-2/SGP2 protein in the influx of extracellular Ca^{2+} into prostatic cells activated to undergo apoptosis.

Apoptosis has been implicated as the molecular mechanism underlying "thymineless" death in other systems (Kyprianou and Isaacs 1989a; Armstrong et al. 1992). This programmed cascade is similar to that activated by androgen ablation in the androgen-responsive human prostatic tumors (Kyprianou et al. 1990) as well as the radiation-induced cell death of androgen-independent human prostate cancer cells (Sklar et al. 1993).

Programmed cell death can also be induced in androgen-independent prostate cancer cells by chemotherapeutic agents that sustain an elevation in the intracellular free calcium level (Martikainen et al. 1991), an induction that does not require these cells to proliferate. Thus, apoptosis is an appropriate therapeutic target for the elimination of androgen-independent prostate cancer cells.

References

Armstrong DK, Isaacs JT, Ottaviano YL, Davidson NE (1992) Programmed cell death in an estrogen-independent human breast cancer cell line, MDA-MB-468. Cancer Res 52:3418–3424

Berns EMMJ, Schnurmans LG, Bolt J, Lamb DJ, Foekens JA, Mulder E (1990) Antiproliferative effects of suramin on androgen responsive tumors. Eur J Cancer 26:470–474

Bettuzi S, Hippakka RA, Gilna P, Liao S (1989) Identification of an androgen-repressed in mRNA in rat ventral prostate is coding for sulphate glycoprotein 2 by cDNA cloning and sequence analysis. Biochem J 257:293–296

Boring CC, Squires TS, Tong T (1993) Cancer statistics. Cancer 43:7–26

Buttyan R, Zaker Z, Lochshin R, Wolgemuth D (1988) Cascade induction of c-fos, c-myc and heat shock 70K transcripts during regression of the rat ventral prostate gland. Mol Endocrinol 2:650–657

Carter BH, Coffey DS (1990) The prostate: an increasing medical problem. Prostate 16:39–48

Cohen, JJ, Duke RC (1984) Glucocorticoid activation of a calcium-dependent endonuclease in thymocyte nuclei leads to cell death. J Immunol 132:38–42

De Clerq E (1979) Suramin: a potent inhibitor of the reverse transcriptase of RNA tumor viruses. Cancer Lett 8:9–22

Eisenberger MA, Reyno LM, Jodrell DI Sinibaldi VJ Tkaczuk KH, Sridhara R, Zuhowski EG, Lowitt MH, Jacobs SC, Egorin MJ (1993) Suramin an active drug for prostate cancer: interim observations in phase I trial. J Natl Cancer Inst 85:611–621

English HF, Kyprianou N, Isaacs JT (1989) Relationship between DNA fragmentation and apoptosis in the programmed cell death in the rat prostate following castration. Prostate 15:233–251

Grem JL (1990) Fluorinated pyrimidines. In: Chabner BA, Collins JM (eds) Cancer chemotherapy. Lippincott, Philadelphia, pp 180–224

Hermens AF, Bentvelzen PAJ (1992) Influence of the H-ras oncogene on radiation responses of rat rhabdomyosarcoma cell line. Cancer Res 52:3073–3082

Isaacs JT (1984) Antagonistic effect of androgen on prostatic cell death. Prostate 5:545–558

Jindal HK, Anderson CN, Davis RG, Vishinanantha JK (1990) Suramin affects DNA synthesis in HELA cells by inhibition of DNA polymerases. Cancer Res 50:7754–7761

Kim JH, Sherwood ER, Sutkowski DM, Lee C, Kozlowski JM (1991) Inhibition of prostatic tumor cell proliferation by suramin: alteration in TFG alpha-mediated autocrine growth regulation and cell cycle distribution. J Urol 146:171–176

Kyprianou N, Isaacs JT (1988) Activation of programmed cell death in the rat ventral prostate after castration. Endocrinology 122:552–562

Kyprianou N, Isaacs JT (1989a) "Thymineless" death in androgen-independent prostatic cancer cells. Biochem Biophys Res Commun 165:73–81

Kyprianou N, Isaacs JT (1989b) Expression of transforming growth factor-β in the rat ventral prostate during castration-induced programmed cell death. Mol Endocrinol 3:1515–1522

Kyprianou N, English HF, Isaacs JT (1988) Activation of a $Ca^{2+} - Mg^{2+}$ dependent endonuclease as an early event in castration-induced prostatic cell death. Prostate 13:103–118

Kyprianou N, English HF, Isaacs JT (1990) Programmed cell death during regression of PC-82 human prostate cancer following androgen ablation. Cancer Res 50:3748–3753

Kyprianou N, Martikainen P, Davis L, English HF, Isaacs JT (1991a) Programmed cell death as a new target for prostatic cancer therapy. Cancer Surv 11:265–277

Kyprianou N, Alexander RB, Isaacs JT (1991b) Activation of programmed cell death by recombinant human tumor necrosis factor plus topoisomerase II targeted drugs in L-929 tumor cells. J Natl Cancer Inst 83:346–349

Kyprianou N, Bains AK, Jacobs SC (1994) Induction of apoptosis in an-
   drogen-independent human prostate cancer cells undergoing thymineless
   death. Prostate 25:12–22
Léger JG, Monpetit MD, Tenniswood MP (1987) Characterization and cloning
   of androgen-repressed mRNA from rat ventral prostate. Biochem Biophys
   Res Commun 147:196–203
Martikainen P, Kyprianou N, Tucker R, Isaacs, JT (1991) Programmed cell
   death of non-proliferating androgen-independent prostate cancer cells.
   Cancer Res 51:4693–4700
McConkey DJ, Nikotera P, Hartzell P, Bellomo G, Wyllie AH, Orrenius S
   (1989) Glucocorticoids activate a suicide process in thymocytes through
   elevation of cytosolic Ca^{2+} concentration. Arch Biochem Biophys
   269:365–370
McKenna WG, Weiss MC, Bakananskas VJ, Sandler H, Kelsten ML, Bioglow
   J, Tuttle SW, Endlich B, Ling CC, Muschel RJ (1990) The role of the H-ras
   oncogene in radiation resistance and metastasis. J Radiat Oncol Biol Phys
   18:849–859
Myers C, Cooper M, Stein C (1992) Suramin: a novel growth factor antagonist
   with activity in hormone-refractory metastatic prostate cancer. J Clin Oncol
   10:881–889
Nicotera P, Hartzell P, Davis G, Orrenius S (1986) The formation of plasma
   membrane blebs in hepatocytes exposed to agents that increase cytosolic
   Ca^{2+} is mediated by the activation of a non-lysosomal proteolytic system.
   Science 209:139–142
Painter RB (1980) The role of DNA damage and repair in cell killing induced
   by ionizing radiation. In: Meyn RE, Withers HR (eds) Radiation biology in
   cancer research. Raven, New York, pp 59–68
Paulson DF, Lin GH, Hinshaw W, Stephani S, Urology and Oncology Research Group (1982) Radical surgery versus radiotherapy for adenocarcinoma of the prostate. J Urol 128:502–504
Peehl D, Wong S, Stamey T (1991) Cytostatic effects of suramin on prostate
   cancer cells cultured from primary tumors. J Urol 145:624–630
Raff MC (1992) Social controls on cell survival and cell death. Nature
   356:397–400
Raghavan D (1988) Non-hormone chemotherapy for prostate cancer: prin-
   ciples of treatment and application to the testing of new drugs. Semin Oncol
   15:371–389
Rago R, Mitchen J, Cheng AL, Oberley T, Wilding G (1991) Disruption of
   cellular energy balance by suramin in intact human prostatic carcinoma
   cells, a likely antiproliferative mechanism. Cancer Res 51:6629–6635

Seewald MJ, Olsen R, Powls G (1989) Suramin blocks intracellular Ca^{2+} release and growth factor-induced increases in cytoplasmic free Ca^{2+} concentration. Cancer Lett 49:107–113

Sensibar JA, Griswold MD, Sylvester SR, Buttyan R, Bardin CN, Cheng CV, Dudek S, Lee C (1992) Prostatic ductal system in rats: regional variation in localization of an androgen-repressed gene product SGP-2. Endocrinology 52:4042–4045

Sklar GN, Eddy HA, Jacobs SC, Kyprianou N (1993) Combined antitumor effect of suramin plus irradiation in human prostate cancer cells: the role of apoptosis. J Urol 150:1526–1532

Stein CA, LaRocca RV, Thomas R, McAtee N, Myers CE (1989) Suramin: an anticancer drug with a unique mechanism of action. J Clin Oncol 7:499–508

Stephens LC, Ang KK, Schulthers TE, Milas L, Meyn R (1991) Apoptosis in irradiated murine tumors. Radiat Res 127:308–316

Sylvester SR, Skinner MK, Griswold MD (1984) A sulphated glycoprotein synthesized by Sertoli cells and by epididymal cells is a component of the sperm membrane. Biol Reprod 31:1087–1101

Tanon I, Janot E (1924) Essai de traitement de la maladie du sommeil, auameroun, par le Bayer 205. Action sur les parasites. Action sur le rein du foie. Ann Paras Hum Comp 2:327–334

Umansky SR, Korol BA, Nelipovich PA (1981) In vivo DNA degradation in thymocytes of γ-irradiated or hydrocortisone-treated rats. Biochem Biophys Acta 655:9–17

Van Steenbrugge GJ, Groen M, Romijin JC, Schroeder F (1984) Biological effects of hormonal treatment regimen on a transplantable human prostate tumor line (PC-82). J Urol 131:812–817

Wyllie AH, Kerr JFR, Currie AR (1980) Cell death: the significance of apoptosis. Int Rev Cytol 68:251–306

Yamada T, Ohyama H (1988) Radiation-induced interphase death of rat thymocytes is internally programmed (apoptosis). Int J Radiat Biol 53:65–75

Yanagihara K, Tsumuraya M (1992). Transforming growth factor β_1 induces apoptotic cell death in cultured human gastric carcinoma cells. Cancer Res 52:4042–4045

Zietman AL, Coen JJ, Shipley WU, Willet CG, Efird JT (1994) Radical radiation therapy in the management of prostatic adenocarcinoma: the initial prostate specific antigen value as a predictor of treatment outcome. J Urol 151:640–645

4 Apoptosis in Relation to Androgen Independence in Experimental and Clinical Prostate Cancer

P. S. Rennie, N. Bruchovsky, K. Akakura, S. L. Goldenberg,
M. Gleave, and L. Sullivan

4.1 Introduction

In the prostate and other androgen-dependent tissues, androgens govern the expression of three types of regulatory genes during the growth and differentiation of stem cells (Bruchovsky et al. 1975). Initiation of DNA synthesis is an example of one type – positive gene regulation by androgens; whereas inhibition of these processes in the presence of a rising titer of hormone typifies another type – negative gene regulation. The third type is exemplified by induction of autophagic lysis (apoptosis) after withdrawal of androgens and involves the expression of a number of androgen-repressed genes (Montpetit et al. 1986; Rennie et al. 1984, 1988). By definition, the third type accounts for androgen dependence. Usually both positive and androgen-repressed types of gene regulation are manifested in androgen-dependent tumors (Bruchovsky et al. 1987). In an androgen-poor environment, most cells in an androgen-dependent tumor undergo apoptosis. However, surviving tumorigenic stem cells irreversibly progress to an androgen-independent condition and give rise to a recurrent tumor with a greatly increased population of androgen-independent stem cells (Bruchovsky et al. 1990).

The capacity to undergo apoptosis is acquired as a feature of differentiation of prostatic cells under the influence of androgens; thus, in the absence of androgens it is impossible for dividing cells to differentiate and to become pre-apoptotic again. This explains why recurrent tumor growth is characterized by androgen independence. In attempting to advert or delay progression to the androgen-independent state, we hypothesized that if tumor cells which survive androgen withdrawal are forced into a normal pathway of differentiation by androgen replacement, then apoptotic potential might be restored. It follows that if androgens are replaced soon after regression of a tumor to a smaller size has taken place, it should be possible to bring about repeated cycles of androgen-stimulated growth, differentiation, and castration-induced regression of tumor.

In the present investigation, the effects of intermittent androgen suppression (i.e., consecutive cycles of androgen ablation and androgen replacement) on the androgen-dependent Shionogi carcinoma were studied in terms of time to androgen independence, change in the proportion of tumorigenic stem cells, and altered regulation of the apoptosis-related gene, TRPM-2 (Tenniswood et al. 1992). In addition, the effects of intermittent androgen suppression on the maintenance of apoptotic potential in several men with prostate cancer was tested (Akakura et al. 1993).

4.2 An Increase in Tumorigenic Stem Cells Is Related to Tumor Progression to Androgen Independence

When the parent androgen-dependent Shionogi carcinoma is transplanted into a male animal, it grows with a doubling time of approximately 24 h (Bruchovsky et al. 1990). Castration of the male host results in almost complete regression of the tumor over succeeding days. Following this, the tumor remains relatively dormant for a period of time after which recurrent androgen-independent growth with a doubling time of 72 h is observed. When the recurrent tumor is transplanted into a normal male host again, the doubling time of the tumor increases, indicating that it is still androgen sensitive. However, castration fails to induce apoptosis and subsequent regression of the tumor, confirming that it is now androgen independent.

To determine whether androgen ablation alters the stem cell composition of the Shionogi carcinoma, we developed an in vivo limiting dilution assay which allowed us to measure the proportion of stem cells at different stages of progression of the Shionogi carcinoma (Bruchovsky et al. 1990). In this assay, single cell suspensions are made from the tumor and counted. A decreasing number of cells ranging from 10^6 to 10^1 is injected subcutaneously into several groups of animals. Over the next 6–8 months, the number of tumor takes is recorded in each group. When the fraction of tumor takes expressed as a percentage is plotted against the number of cells implanted, a sigmoidal curve relationship is obtained. At the point at which the fraction of tumor takes equals 37%, there is on average one stem cell per inoculum. Injection of tumor cells into male animals yields an estimate of the proportion of total stem cells in the overall tumor cell population. In the parent tumor, this proportion is one stem cell per 4000 tumor cells. In the recurrent tumor, the proportion is one stem cell per 200 tumor cells. The difference between the parent tumor and the recurrent tumor is consistent with a 20-fold increase in the proportion of total stem cells in a tumor.

4.3 The Effects of Intermittent Androgen Suppression on the Stem Cell Population of the Shionogi Carcinoma

The effects of intermittent androgen suppression on progression of Shionogi carcinoma was investigated using the Shionogi tumor model (Akakura et al. 1993). The parent androgen-dependent tumor was transplanted into a succession of male mice, each of which was castrated when the estimated tumor weight became about 3 g. After the tumor had regressed to 30% of the original weight, it was transplanted into another noncastrated male. This cycle of transplantation and castration-induced apoptosis was repeated successfully three times before growth became androgen independent during the fourth cycle. With this intermittent androgen suppression regimen, the time required to progress to androgen independence is about 120 days. By comparison, if the tumor is treated by one-time castration only, regression of the tumor is followed by recurrence after an interval of about 50 days.

With respect to the stem cell population, the proportion in each cycle-specific tumor remains fairly constant at about one stem cell per 4000 tumor cells in cycle-1, cycle-2, and cycle-3 tumors. In the cycle-4 tumor, the proportion of total stem cells increases to one stem cell per 310 tumor cells, which is virtually identical to the proportion in a recurrent androgen-independent tumor.

The proportion of the androgen-independent stem cells in each of the cycle-specific tumors was detected using the in vivo limiting dilution assay with female animals. These results indicated that the proportion of androgen-independent stem cells increases steadily in each cycle of therapy until the stem cell composition is identical to that in the recurrent tumor. In the cycle-1 tumor at the start of therapy, the androgen-independent stem cells contribute less than 1% to the total stem cell population. In the cycle-4 and recurrent tumors, the androgen-independent stem cells form 50% of the total stem cell compartment. Therefore, it appears that a tumor becomes androgen independent when one half of the total stem cell compartment is populated by androgen-independent stem cells. This end-result occurs more quickly in a single step after one-time castration; with intermittent androgen suppression, the same end result develops but over a longer time frame. Thus, tumor progression is delayed but not prevented by intermittent androgen suppression.

4.4 Constitutive Overexpression of TRPM-2 (Clusterin) Is Related to Tumor Progression

In searching for factors which might account for the dramatic increase in the proportion of stem cells during tumor progression, we focused our attention on the expression of the TRPM-2 gene, which is closely associated with active cell death (Montpetit et al. 1986; Bettuzzi et al. 1989; Rennie et al. 1988; Buttyan et al. 1989). The TRPM-2 gene is androgen repressed and only becomes active after androgens are withdrawn. A 2.0-kb band representing the mRNA transcript for TRPM-2 is detected by northern hybridization analysis. Using the intermittent androgen suppression protocol in the Shionogi model, the following pattern is observed. In the parent cycle-1 tumor before castration, very little transcript is detected; after castration, there is a marked increase in the intensity of the band, indicating upregulation in the absence of androgen. With subsequent cycles of transplantation and castration, there is a gradual loss of androgenic repression of TRPM-2 expression, such that in the cycle-4 tumor there is very little difference in the amount of transcript before and after castration. This coincides with the development of androgen independence. Thus, constitutive overexpression of the TRPM-2 gene appears to be closely related to tumor progression and the loss of apoptotic potential.

4.5 Nuclear Localization of Clusterin and Loss of Apoptotic Potential

Using an anti-clusterin antibody, we next studied the appearance and localization of clusterin, the protein product of the TRPM-2 gene, in the Shionogi carcinoma during intermittent androgen suppression. The parent Shionogi carcinoma before castration revealed no immunohistochemical staining for clusterin. After castration, the cytoplasm of regressing cells is intensely stained with anti-clusterin antibody. In the cycle-2 tumors before castration, weak diffuse immunostaining of cytoplasm and staining of a small number of nuclei is observed. In the cycle-4 tumors in noncastrated hosts, the staining of cytoplasm is also diffuse but more nuclear localization of clusterin is demonstrated. An impressive feature of the recurrent tumor is intense staining of many nuclei with the anti-clusterin antibody. Thus, tumor progression is linked to the paradoxical overexpression and deregulation of clusterin, the anticytolytic properties of which could limit or prevent the onset of apoptosis and thereby account for androgen independence.

4.6 The Effects of Intermittent Androgen Suppression on Prostate-Specific Antigen Secretion by LNCaP Cells

The in vivo LNCaP tumor model was also used to test the effects of intermittent androgen suppression. With a supporting matrix, this tumor grows in male nude mice and secretes prostate-specific antigen (PSA) into the serum (Gleave et al. 1992). After one-time castration, the serum PSA initially decreases approximately fivefold and then slowly increases at a constant rate. If testosterone is replaced in the form of a pellet 7 days after castration, the serum PSA increases more rapidly. When the pellet is removed, the serum PSA falls to a lower level. Intermittent testosterone withdrawal and replacement in this system prevents the baseline serum PSA from increasing as rapidly as after one-time castration and delays the onset of androgen-independent regulation of the PSA gene.

4.7 Intermittent Androgen Suppression for the Treatment of Prostate Cancer

Intermittent androgen suppression for the treatment of human prostate cancer is achievable owing to the reversibility of agents now available which lower the effective intranuclear concentration of androgens. Reversible androgen withdrawal therapy is usually accomplished by medical castration (e.g., luteinizing hormone-releasing hormone, LHRH agonists) combined with antiandrogens. The pharmaceutical agents for intermittent androgen suppression should fulfill certain criteria so that the necessary long-term clinical and laboratory monitoring are logical and informative for both the patient and physicians. These criteria are as follows: (a) rapid suppression of androgens; (b) reversibility of action; (c) absence of flare reaction; (d) low toxicity; and (f) dose dependence.

Owing to several positive features, "lead-in" therapy plus combined androgen blockade is a strong first option for intermittent androgen suppression. Our application of this protocol uses diethylstilbestrol (DES) and cyproterone acetate for lead-in therapy and an LHRH agonist with cyproterone acetate for androgen blockade. Lead-in therapy involves administration of cyproterone acetate (100 mg/day) and DES (0.1 mg/day) for 4 weeks – in most patients, this brings serum testosterone into the castrate range within 3–7 days and eliminates the flare reaction seen with the use of LHRH agonists. Goserelin acetate (3.6 mg) (or leuprolide acetate) is started after 4 weeks of lead-in therapy and administered every 4 weeks thereafter. DES is halted after 4 weeks of lead-in therapy but cyproterone acetate is continued to prevent hot flushes. The dose of cyproterone acetate may be increased by 50 mg/day to a total of 300 mg/day. Cyproterone acetate administration continues until the end of 36 weeks or 9 months. Serum testosterone recovers to the normal range 8–16 weeks after cessation of therapy. This treatment regimen features rapid onset of action, no risk of flare reaction, low incidence of hot flushes, few side effects other than loss of libido and potency, a dose-responsive effect, reversibility, and excellent compliance.

In the ideal case, the patient is on therapy for an overall period of 8–9 months. The serum PSA is observed to decrease to the lower limit of detection, 0.2 ng/ml, after 6 months of therapy. The same value is obtained after 8 months of therapy, indicating that the serum PSA has reached a stable nadir in the normal range. Therapy is then interrupted after 9 months, marking the start of the off-treatment period. The serum testosterone levels are restored to normal within 8–14 weeks. Unlike the rate of recovery of serum testosterone, the increase in serum PSA can be quite slow allowing a long no-treatment period of 10 months or more (Akakura et al. 1993). Although the majority of patients whom we have treated with intermittent androgen suppression have completed just one cycle, many have completed two or three cycles and one patient has responded five times over a period of 55 months. The tumor in this patient became androgen independent after the fifth cycle as shown by an increase in serum levels of PSA despite continuing suppression of serum testosterone (Akakura et al. 1993).

4.8 Serum PSA as a Marker of Response and Prognosis

In about 70% of men with advanced prostate cancer, serum PSA reverts
to normal during the first 8 months of androgen suppression and criteria
for interruption of therapy are met. In the remaining 30%, serum PSA
decreases temporarily and then increases; if a plateau is reached, it is
short-lived or stabilizes outside of the normal range. This group is not
eligible for interruption of therapy and other treatment in addition to
androgen suppression should be considered. It has been observed that
the rate of normalization is faster between zero time and 10 weeks than
in the succeeding interval between 10 and 32 weeks of initial treatment
(Bruchovsky et al. 1993). From this point onward, the serum PSA levels
remain abnormal in about 30% of patients, as described above. These
results indicate that PSA response is most likely to be observed in the
first 32 weeks of therapy, with little chance of treatment producing a
normal result beyond this time.

4.9 Effect of Intermittent Androgen Suppression
on Survival

It is unknown whether intermittent androgen suppression and its effects
on tumor progression alters survival in a beneficial or adverse way,
although the possibility of an improved outcome has been considered by
others (Klotz et al. 1986). Our experience with this therapy has indicated no unfavorable effects on survival. Up until now, the concept of
intermittent androgen suppression has not been tested in a randomized
clinical trial but formal studies of this nature will be started in the next
few months by the Southwest Oncology Group in the United States. The
results of intermittent versus continuous androgen suppression in patients in whom the serum PSA is lowered into the normal range by
initial therapy will be compared. Our observations on the results of
intermittent androgen suppression in a small group of patients are summarized in Table 1.

Table 1. Preliminary clinical data using intermittent androgen suppression

Patients	35 (3 with stage A2 prostate cancer, 5 with stage B2, 11 with stage C, and 16 with stage D)
Mean follow-up time	30 months (range 6–79)
Mean pretreatment PSA level	174 mg/l
Mean time to PSA nadir	5 months (range 1–14)
Mean time to testosterone recovery	2 months
Mean time off treatment	5 months (range 1–13)
Mean time to androgen independence	43 months (2–5 cycles)
Percentage of no-treatment time	31 % (range 13 %–77 %) of total time

PSA, prostate-specific antigen.

4.10 Advantages of Intermittent Androgen Suppression

Whether intermittent androgen suppression enhances progression-free survival or overall survival will be determined in future randomized clinical trials. Our results with the Shionogi tumor model suggest that intermittent androgen suppression delays progression to androgen inde-

Table 2. Advantages of intermittent androgen suppression

Preservation of apoptotic potential of tumor
Possible delay in progression to androgen independence
Improved quality of life
  • Recovery of libido and potency
  • Greater sense of well-being
  • Less reduction of bone mass
Reduced expense of treatment
New options for chemotherapy
  • At time of PSA nadir (to eliminate stem cells)
  • During recovery of testicular function (to eliminate androgen-sensitive dividing cells)
More potential applications in treating all stages of prostate cancer

PSA, prostate-specific antigen.

pendence but does not prevent it. However, intermittent androgen suppression does provide for improved quality of life characterized by a
markedly greater sense of well-being during off-treatment periods. The
overall benefits of intermittent androgen suppression are summarized in
Table 2.

4.11 Conclusions

Androgen suppression is an effective approach to the treatment of
advanced prostate cancer and can be accomplished by surgical orchiectomy or by employing agents which inhibit the synthesis and peripheral
action of testosterone. Our results with the Shionogi carcinoma indicate
that androgen ablation promotes an increase in the proportion of tumorigenic stem cells, as well as deregulation of TRPM-2 expression and
nuclear localization of its protein product clusterin. The anomalous
presence of clusterin in the nucleus may serve to inhibit early events in
the apoptotic process and thereby foster the generation and outgrowth of
androgen-independent stem cells in an androgen-depleted environment.

In both experimental and clinical carcinomas, apoptosis can be induced multiple times using successive cycles of androgen withdrawal
and replacement (intermittent androgen suppression). The preservation
of this capacity for apoptosis by intermittent androgen suppression is
probably achieved through androgen-induced differentiation of stem
cells. However, in the Shionogi carcinoma and the LNCaP tumor
model, progression to the androgen-independent state can be retarded
but not prevented by intermittent androgen suppression (Akakura et al.
1993; Gleave, unpublished).

Under such conditions of deliberate and gradually increasing hormonal stimulation, the androgen-dependent (apoptotic) state of the
tumor will be restored, thus setting the stage for another response to
androgen deprivation. Potential benefits of intermittent androgen suppression include an improved quality of life with recovery of sexual
function, prolongation of the androgen-dependent state of the tumor,
reduced cost of treatment, and possible application to the management
of the disease in earlier stages.

Only those patients in whom serum PSA levels are normalized by
initial androgen deprivation should be considered as candidates for

intermittent therapy. The failure of serum PSA to decrease to a stable level within the normal range during the first 8 months of androgen suppression is a sign of early progression to androgen independence and is associated with a poor prognosis. Whether intermittent androgen suppression and its effects on tumor progression alter survival in a beneficial or adverse way is unknown; however, no unfavorable effects have been suggested by clinical experience with such therapy to date.

References

Akakura K, Bruchovsky N, Goldenberg SL, Rennie PS, Buckley AR, Sullivan LD (1993) Effects of intermittent androgen suppression on androgen-dependent tumors. Cancer 71:2782–2790

Bettuzzi S, Hiipakka RA, Gilna P, Liao S (1989) Identification of an androgen-repressed mRNA in rat ventral prostate as coding for sulphated glycoprotein 2 by cDNA cloning and sequence analysis. Biochem J 257:293–296

Bruchovsky N, Lesser B, Van Doorn E, Craven S (1975) Hormonal effects on cell proliferation in rat prostate. Vitam Horm 33:61–102

Bruchovsky N, Brown EM, Coppin CM, Goldenberg SL, Le Riche JC, Murray NC, Rennie PS (1987) The endocrinology and treatment of prostate tumor progression. In: Coffey DS, Bruchovsky N, Gardner WA Jr, Resnick MI, Karr JP (eds) Current concepts and approaches to the study of prostate cancer. Liss, New York, pp 347–387

Bruchovsky N, Rennie PS, Coldman AJ, Goldenberg SL, To M, Lawson D (1990) Effects of androgen withdrawal on the stem cell composition of the Shionogi carcinoma. Cancer Res 50:2275–2282

Bruchovsky N, Goldenberg SL, Akakura K, Rennie PS (1993) LHRH agonists in prostate cancer: elimination of flare reaction by pretreatment with cyproterone acetate and low-dose diethylstilbestrol. Cancer 72:1685–1691

Buttyan R, Olsson CA, Pintar J, Chang C, Bandyk M, Ng P, Sawczuk IS (1989) Induction of the TRPM-2 gene in cells undergoing programmed death. Mol Cell Biol 9:3473–3481

Gleave M, Hsieh J-T, Wu HC, von Eschenbach AC, Chung LWK (1992) Serum prostate specific antigen levels in mice bearing human prostate LNCaP tumors are determined by tumor volume and endocrine and growth factors. Cancer Res 52:1598–1605

Klotz LH, Kerr HW, Morse MJ, Whitmore WF Jr (1986) Intermittent endocrine therapy for advanced prostate cancer. Cancer 58:2546–2550

Montpetit ML, Lawless KR, Tenniswood M (1986) Androgen-repressed messages in the rat ventral prostate. Prostate 8:25–36

Rennie PS, Bouffard R, Bruchovsky N, Cheng H (1984) Increased activity of plasminogen activators during involution of the rat ventral prostate. Biochem J 221:171–178

Rennie PS, Bruchovsky N, Buttyan R, Benson M, Cheng H (1988) Gene expression during the early phases of regression of the androgen-dependent Shionogi mouse mammary carcinoma. Cancer Res 48:6309–6312

Tenniswood MP, Guenette RS, Lakins J, Mooibroek M, Wong P, Welsh J-E (1992) Active cell death in hormone-dependent tissues. Cancer Metastasis Rev 11:197–220

5 Active Cell Death and Cancer

W. Bursch, B. Grasl-Kraupp, A. Ellinger, L. Török, H. Kienzl,
L. Müllauer, and R. Schulte-Hermann

5.1 Introduction

Cancer development is a complex, multistage process in which failure
of cell differentiation and proliferation control, for example, after activation of oncogenes, is considered to be an important causative factor.
Relatively little attention has been paid to the role of cell death in cancer
development, probably because for a long time in toxicology cell death
was regarded as a passive degenerative phenomenon of secondary importance for the regulation of cell number in tissues. However, cell
death is a biological phenomenon as important as cell proliferation. It
serves to shape the final form of organisms during embryonic development and metamorphosis (Ellis et al. 1991; Glücksmann 1951; Lock-

shin and Williams 1964; Saunders 1966) and it counterbalances cell generation in adult tissues. Cell death is also a significant result of tissue damage and cause of disease (Farber et al. 1972; Kerr et al. 1972; Popper and Keppler 1986; Wyllie et al. 1980). Recently, new concepts have emerged related to the different types of cell death occurring in this wide variety of circumstances. In 1972 J. Kerr, A. Wyllie, and A. Currie proposed a classification of cell death into two broad categories. They introduced the concept of apoptosis to describe a form of active self-destruction of a cell which is under the control of the growth-regulating network and which – functionally speaking the opposite of mitosis – serves to downregulate the cell number in tissues (Kerr et al. 1972; Wyllie et al. 1980). According to this proposal, necrosis results from violent environmental perturbation, leading to rapid incapacitation of major cell functions (gene expression, ATP synthesis, membrane potential) and to the collapse of internal homeostasis. This concept eventually helped to elucidate the role of cell death in a variety of (patho)physiological states (Bursch et al. 1992; Fesus 1991; Korsmeyer 1992). Thus, during the past decades increasing evidence has accumulated showing that disturbance of apoptosis is involved in the pathogenesis of tumors (Bursch et al. 1992; Korsmeyer 1992; Sarraf and Bowen 1988; Vaux et al. 1988). Evidence for apoptosis was also found during regression of hormone-dependent tumors (Bursch 1994; Bursch et al. 1991; Gullino 1980; Kerr et al. 1972; Kyprianou et al. 1990, 1991; Lanzerotti and Gullino 1972; Szende et al. 1989, 1990). Thereby, the concept of apoptosis stimulated the development of new strategies for cancer treatment.

5.2 Morphology of Active Cell Death

Apoptosis occurs through a series of morphologically distinct alterations which comprise shrinkage of cytoplasm, condensation of chromatin, and fragmentation of the affected cell into membrane-bound "apoptotic bodies" (Kerr et al. 1972). The skrinkage of cytoplasm in the early stages of apoptosis does not involve lysosomal digestion of cytoplasmic constituents. Another basic feature of apoptosis under many circumstances is that the dying cell and its fragments are secluded by intact membranes until the final stage of digestion after phagocytosis. The rapid removal of apoptotic cells through phagocytosis is facilitated by the expression of surface molecules such as lectin binding sites (Wyllie et al. 1984), vitronectin receptor (Savill et al. 1990), phosphatidyl-serine (Fadok et al. 1992), or asialoglycoprotein receptor (Dini et al. 1992). Thereby, liberation of potentially harmful substances such as DNA, antigens, and eicosanoids and activation of macrophages with formation of oxygen radicals and inflammation may be avoided. Necrosis, in contrast, is associated with membrane lysis and inflammation (Popper and Keppler 1986), which may cause secondary tissue damage. Thus, from a teleological point of view, apoptosis appears to be more advantageous for removal of injured cells than necrosis.

While in recent years apoptosis has attracted increasing attention, other types of *active* or "programmed" cell death have been recognized with a morphology clearly different from apoptosis. Schweichel and Merker (1973) observed in embryonal tissues a type of cell death during which early activation of lysosomes and formation of autophagic vacuoles is the predominant feature. There have been similar findings in insect tissue during metamorphosis (Lockshin and Beaulaton 1974; Lockshin and Williams 1965; Zakeri et al. 1993) and more recently in neuronal tissue (Clarke 1990). The lysosomal or type II cell death is probably involved in regression of mammary tumors after withdrawal of estrogens (Lanzerotti and Gullino 1972) and has been observed in the human mammary cancer cell line MCF7 after antiestrogen treatment (Bursch et al., submitted). In the present paper the term "active cell death" is used to encompass its different morphologial subtypes, since apoptosis originally was described as a type of cell death *without primary* lysosomal digestion of cellular material (Kerr et al. 1972).

5.3 DNA Degradation During Active Cell Death

An early metabolic change associated with apoptosis is considered to be the activation of a nonlysosomal endonuclease, which cuts chromatin into oligonucleosomes (Arends et al. 1990; Wyllie 1980). The resulting chromatin fragmentation yields a characteristic ladder pattern after gel electrophoresis and is frequently used to detect apoptosis. However, morphologically proven apoptosis in hepatocytes and other epithelial cells can occur without DNA fragmentation into oligonucleosomes (Oberhammer et al. 1992). Recent evidence suggests that during chromatin condensation large DNA fragments of 50 and 300 kbp are formed, proabably by activation of topoisomerase II (Filipski et al. 1990; Brown et al. 1993; Oberhammer et al. 1993b; Walker et al. 1993). Further degradation of chromatin during later stages of apoptosis may yield DNA fragments of (oligo)nucleosomal or irregular size (depending on the cell type). It is of interest that DNA fragmentations can be demonstrated on tissue sections by nick translation in situ (Oberhammer et al. 1993c) but it should be noted that these tests are not specific for apoptosis or any type of active cell death. Furthermore, endonuclease is constitutively present in intact, nonapoptotic nuclei and may also be activated by mechanisms not related to apoptosis but rather to necrosis (G.M. Cohen et al. 1992, Collins et al. 1992) or in lytic cells postmortem (Martz and Howell 1989). Therefore endonuclease activation and detection of the "DNA ladder" should not be considered as a specific marker of apoptosis.

5.4 Active Cell Death in Normal Tissue

Active cell death may serve to eliminate excessive cells from hyperplastic but otherwise normal organs, as exemplified by rat liver growth and regression after treatment with xenobiotic compounds such as phenobarbital, cyproterone acetate (CPA), and others (Bursch et al. 1984; Grasl-Kraupp et al. 1993; Schulte-Hermann et al. 1990). Repeated administration of high doses of CPA produces massive hyperplasia of rat liver, which after cessation of CPA treatment partially regresses within a matter of a few days (Bursch et al. 1984, 1986). The regression of liver hyperplasia is brought about by apoptosis as demonstrated by morphological and histochemical means (Bursch et al. 1985). An important functional property is that apoptosis can be triggered by withdrawal of short-lived mitogens such as CPA. On the other hand, apoptosis is inhibited by tissue-specific mitogens (Bursch et al. 1984, 1992; Isaacs 1984; Levi-Montalcini 1987; Wyllie et al. 1980). Thus, the tissue levels of mitogens (growth factors) may determine a delicate balance between the rates of cell division and death and thereby regulate homeostasis of cell number in tissues.

5.5 Kinetics of Active Cell Death

Inhibition by liver mitogens could be also used to estimate the duration of the histologically visible stages of apoptosis. For this purpose the time course of elimination of apoptotic bodies in a regressing hyperplastic liver after CPA treatment was closely followed, and an average duration of 3 h was found (Bursch et al. 1990). A similar duration of the histological stages of apoptosis was found in preneoplastic lesions of rat liver (Bursch et al. 1990) and in an estrogen-dependent kidney tumor (Bursch et al. 1991). These findings allowed the calculation of the rate of apoptosis from histological counts, which was about 0.5%/h in normal liver and 5%/h in preneoplastic liver foci (Bursch et al. 1990).

5.6 Genes Regulating Active Cell Death

Recent studies on oncogene expression have revealed exciting insights into the genetic control of apoptosis. In general, there appear to be genes which favor active cell death, such as myc, bax and p53; others are inhibitory, e.g., bcl-2 in B-cells or ced-9 in the nematode *Caenorhabditis elegans* (Bissonette et al. 1992; Buttyan et al. 1988; Evan et al. 1992; Fanidi et al. 1992; Hockenberry et al. 1990; Hengartner et al. 1992; Oltvai et al. 1993; Vaux ct al. 1988; Yonish-Roauch et al. 1991). Early growth response genes such as myc may prime some normal and transformed cells for apoptosis. The decision, however, as to whether such cells enter the apoptosis versus proliferation pathway depends on cooperative signals which may be provided by bcl-2 (inhibition, see above) or p53. In thymocytes p53 seems to be involved in apoptosis induction by genotoxic chemicals and radiation. In contrast, glucocorticoid-induced thymocyte apoptosis does not appear to depend on p53 (Clarke et al. 1993; Lowe et al. 1993).

We have studied the gene coding for testosterone-repressed prostate message-2 (TRPM-2) in some detail. The TRPM-2 gene was originally discovered during regression by apoptosis of prostate after castration (Montpetit et al. 1986; Léger et al. 1988). By northern blot analysis it was later also found to be expressed in the liver during involution of mitogen-induced hyperplasia (Bursch et al. 1994). In situ hybridization studies, however, have revealed that the enhanced TRPM-2 expression during liver involution is not specifically associated with individual cells undergoing apoptosis but rather is found in almost all hepatocytes. Moreover, TRPM-2 was also found to be expressed during the stages of rat liver necrosis and regeneration after intoxication with CCl_4. Sequence analysis of TRPM-2 has revealed that TRPM-2 is homologous to clusterin, complement lysis inhibitor, and other proteins (Jenne and Tschopp 1992). Therefore, it was concluded that the role of this gene may be a more general one, namely, to protect cells from complement-induced lysis during tissue remodeling as it occurs for example during liver growth and regression (Bursch et al. 1994; Kirszbaum et al. 1989; Tenniswood et al. 1992).

5.7 Factors Regulating Active Cell Death

Factors involved in the regulation of apoptosis include certain hormone agonists and antagonists (see below), tumor necrosis factor (TNF)-α, and transforming growth factor (TGF)-β_1 (Bellomo et al. 1992; Laster et al. 1988; Ogasawara et al. 1993; Rotello et al. 1991; Shinagawa et al. 1991). Our in vivo studies have shown that apoptotic hepatocytes in normal and preneoplastic liver exhibit a positive immunostaining for TGF-β_1. The staining is much stronger with antibodies recognizing the latency-associated protein (LAP; dimer of the pro-region noncovalently associated with the mature region) rather than the mature peptide itself (Bursch et al. 1993). This may be explained by the biological half-life of the pre-form of TGF-β_1, which is about 2 h, whereas that of the mature TGF-β_1 is only 2 min. Thus, once mature TGF-β_1 is released from the precursor by proteolytic cleavage, its rapid degradation may result in low levels in the cells that may be insufficient for detection under the experimental conditions used (Bursch et al. 1993). Further support for a regulatory role of TGF-β_1 in apoptosis was provided by the finding that TGF-β_1 can induce apoptosis of hepatocytes, both in vitro and in vivo. Interestingly, CPA pretreatment and TGF-β_1 exhibited a synergistic effect on the induction of apoptosis in vivo. This observation suggests that events linked to CPA treatment prime hepatocytes to enter apoptosis after TGF-β_1 (Oberhammer et al. 1993a).

Interestingly, some seemingly intact hepatocytes exhibited a positive immunostaining for pre-TGF-β_1, too. Closer analysis revealed that their incidence correlated with the incidence of apoptotic bodies and, further-more, that the pre-TGF-β_1 positive, intact hepatocytes were about 15% smaller than their negative counterparts (Bursch et al. 1993). These observations suggest that hepatocytes preparing for apoptosis can be detected by the antibody against pre-TGF-β_1 a few hours before chromatin condensation occurs (Bursch et al. 1992, 1993). Furthermore, the TGF-β_1 precursor does not appear to be present in necrotic cells and therefore, immunostaining for this protein may also became a useful marker to descriminate apoptosis from necrosis (Bursch et al. 1993).

Another member of the TGF-β family of polypeptides is activin. This peptide, like TGF-β_1, was found to induce apoptosis in intact rat liver as well as in isolated hepatocytes (Schwall et al. 1993). These studies support the view that TGF-β_1 and related peptides are involved in the regulation of the balance between cell proliferation and cell death in the liver.

5.8 Nutrition and Active Cell Death

The feeding regimen has been found to have a profound influence on the occurrence of apoptosis in normal and hyperplastic liver. Prolonged fasting or severe starvation increased the rate of apoptosis in rat liver (Schulte Hermann et al. 1988; Grasl-Kraupp et al. 1994). On the other hand, food uptake exerted an inhibitory effect on the occurrence of apoptosis during the regression of liver hyperplasia. The extent as well as the kinetics of the food-mediated inhibition of apoptosis were almost identical to that observed after mitogen treatment. Furthermore, inhibition of apoptosis was found to occur not only after consumption of the complete diet, but also after feeding of isocaloric amounts of either pure carbohydrate, protein, or fat. This observation suggests that energy uptake per se rather than specific dietary components are effective. Furthermore, the cyclic eating behavior of rats seems to result in a diurnal rhythm of cell turnover with a peak of apoptosis expected to occur shortly before feeding, while the peak of cell proliferation is known to occur about 12 h later (Schulte-Hermann 1977).

5.9 Active Cell Death in Cancer Development

Apoptosis has been found to modulate the stages of initiation, promotion, and progression in the stepwise formation of cancer in the liver and other organs. In the liver putative initiated cells can be detected as single cells or foci of cells by immunocytochemical and histochemical means. The growth rate of these lesions is considered to be one of the rate-limiting steps in multistage carcinogensis (Farber and Cameron 1980; Pitot and Sirica 1980; Schulte-Hermann 1985). An early observation was that putative preneoplastic foci exhibited much higher rates of cell proliferation than normal liver but showed almost no net growth during several months (Schulte-Hermann et al. 1983). Nongenotoxic carcinogens such as α-hexachlorocyclohexane or phenobarbital (PB) dramatically enhanced foci growth, but surprisingly without persistent significant enhancement of cell proliferation. These paradoxes were resolved by the discovery of apoptosis in foci (Bursch et al. 1984). Apoptosis can efficiently counterbalance the enhanced rate of cell replication in liver foci. PB (and other tumor promoters) can inhibit apoptosis in liver foci and thus favor their growth and the development of frank neoplasia (Bursch et al. 1984, Schulte-Hermann et al. 1990). Furthermore, the rate of apoptosis is inversely correlated with the stability of the expression of the altered phenotype (Schulte-Hermann et al. 1990). Obviously, PB and similar tumor promoters act like trophic hormones or mitogens (see above). The occurrence of apoptosis in liver foci has subsequently been confirmed and extended by other groups (Columbano et al. 1984; Garcea et al. 1989).

The occurrence of apoptosis also has other important implications:
1. It may explain the high probability of extinction of initiated cells as predicted by biologically based mathematical models (80%–90% of initiated cells were calculated not to develop into observable lesions; Moolgavkar et al. 1990; Luebeck et al. 1991). Experimental support for these calculations was provided by the fate of initiated liver cells as visualized by glutathione-S-transferase (GST-P). After a single necrogenic dose of N-nitrosomorpholine, their number steeply increased during regenerative hepatocellular proliferation. After a plateau phase their number decreased to approximately 20% of the peak value. The disappearance of single GST-P positive cells is clearly not due to cell division resulting in focal cell populations

consisting of two and more cells (Grasl-Kraupp, Wagner, Ruttkay-Nedecky, Schulte-Hermann, manuscript in preparation). Although it is not yet proven that the GST-P positive cells are truely initiated and that they disappear only through apoptosis, these findings suggest that initiation as indicated by the occurrence of cells with an altered phenotype is not as stable as is usually assumed.

2. Cessation of PB treatment after a prolonged period of time produced a striking increase in apoptosis in liver foci (Bursch et al. 1984), providing an explanation of how tumor promotion may be *reversible*. Apoptosis in liver foci was also found to stimulated by *S*-adenosyl-L-methionine, resulting in *prevention* of hepatocarcinogenesis (Garcea et al. 1989; Pascale et al. 1992).

3. In view of our findings that food restriction suppresses cell proliferation and concomitantly enhances apoptosis of normal hepatocytes (see above) we have determined that this holds true for preneoplastic liver cells as well. For this purpose, aged rats that had developed a substantial number of "spontaneous" preneoplastic liver foci (i.e., no genotoxic initiator was administered; Grasl-Kraupp et al. 1991) were subjected to food restriction for 3 months. The fasting regimen induces in normal liver and in putative preneoplastic foci a decrease of DNA replication and an increase of apoptosis, both effects being much more pronounced in foci than in normal liver. As a result, at the end of the fasting period the number of normal liver cells (indicated by total liver DNA) had decreased by 15%, while foci had declined by 90%. Subsequent treatment with the liver tumor promoter nafenopin demonstrated findings that many initiated clones had been eliminated (Grasl-Kraupp et al. 1994). This suggests that fasting shifts the balance between cell proliferation and apoptosis in both normal and preneoplastic liver tissue in favor of apoptosis, but more strongly in foci. This observation may provide a new explanation for the old finding in both experimental animals and human populations that restricted feeding can lower cancer rates while overfeeding may enhance them.

4. It is also noteworthy that in the absence of treatment with a tumor promoter, preneoplastic foci exhibited a slow growth in later stages (after about 7 months). Conceivably, the high cell turnover in the early foci may result in selection of preneoplastic cell populations that gradually evade the apoptotic defense mechanism.

Suppression of apoptosis appears to be a pathogenic mechanism of general importance as it occurs not only in the liver, but also in other organs. Block of apoptotic elimination of B cells as a result of overexpression of the oncogene bcl-2 is involved in the pathogenesis of Burkitt lymphoma (Fanidi et al. 1992; Korsmeyer 1992; Vaux et al. 1988). Studies on breast biopsies of premenopausal women suggest that a decreased rate in apoptotic elimination of breast epithelial cells is associated with the occurrence of fibrocytic change and increased risk of development of carcinoma (Allan et al. 1992).

5.10 Active Cell Death and Cancer Therapy

Apoptosis also occurs at later stages of carcinoma development such as in neoplastic nodules and also quite frequently in untreated experimental and human tumors (Sarraf and Bowen 1986, 1988; Wyllie 1985). It is important to realize that, in neoplasia, cell proliferation and active cell death still may depend on growth and survival factors. For example, an estrogen-dependent kidney tumor shows rapid growth when treated with diethylstilbestrol (DES); withdrawal of DES causes tumor regression; and retreatment induced tumor growth again (Bursch et al. 1991). During the regression period there is low high mitotic activity and numerous apoptoses, whereas the opposite occurs during the growth period. It should be noted that large necrotic areas are also present in these tumors, the incidence of which, however, do not change in the presence or absence of DES. Thus the actual growth rate of the tumor depends primarily on the ratio between proliferation and apoptosis. In chemically induced pancreatic cancer in hamsters, apoptosis and tumor regression can be induced by the luteinizing hormone-releasing hormone (LH-RH) analog D-Trp-6-LH-RH, the LH-RH antagonist SB-75, and the somatostatin analog RC-160. The combination of these compounds or of either compound with 5-fluorouracil has been found to increase the efficacy of the therapy of pancreatic tumors (Szende et al. 1989, 1990; Szepeshazi et al. 1991). SB-75 has also been shown to be effective in human prostate carcinoma (PC-82) xenografts in nude mice (Redding et al. 1992). Furthermore, progesterone antagonists exerted a tumor-inhibiting effect in various hormone-dependent mammary tumor models; the antitumor activity of antiprogestins is considered to result from induction of terminal differentiation, leading to apoptosis of the tumor cells (Michna et al. 1992). In vivo studies with the human prostate PC-82 and the mammary cancer cell line MCF-7 revealed that ablation of androgen and estrogen, respectively, induced apoptosis and tumor regression (Kyprianou et al. 1990, 1991).

Likewise, in MCF-7 cell cultures tamoxifen and other antiestrogens induced both a depression of DNA synthesis and an increase in cell death (Bardon et al. 1987; Bursch et al., submitted; Wärri et al. 1993). To what extent induction of cell death may contribute to the preventive effect of tamoxifen against mammary cancer development in patients at risk remains to be elucidated. Conceivably, this could result from anti-promotion by the antagonist via induction of apoptosis in cancer pres-tages. Apoptosis may also have general importance in the therapy of human lymphatic leukemia as suggested by glucocorticoid-induced apoptosis of normal and transformed hematological cells (J.J. Cohen et al. 1992; Fesus 1991; Korsmeyer 1992; MacDonald and Lees 1990; Wyllie et al. 1980) as well as by the induction of death of these cells through cytostatic drugs (Eastman 1990).

To date it is not clear whether TGF-β_1 plays a significant role in tumor therapy. TGF-β_1 has been found to induce apoptosis in cultured human hepatoma and gastric carcinoma cells (Fukuda et al. 1993; Lin and Chou 1992; Yanagihara and Tsumueaya 1992). Some tumor cells produce TGF-β_1 but appear to be resistant to its growth inhibitory effects (Valverius et al. 1989). Breast tumors of patients growing des-pite tamoxifen treatment show high levels of TGF-β_1 mRNA levels and clinically insignificant amounts of estrogen receptors (Thompson et al. 1991). Likewise, TGF-β_1 did not affect the growth of the estrogen receptor-negative human breast cancer line MDA-MB-231 in vivo either (Zugmaier et al. 1991). In our studies with MCF-7 cells, TGF-β_1 did not affect either proliferation or cell death. Furthermore, TGF-β_1 alone or in combination with tamoxifen failed to cause regression of dimethylbenzanthracene (DMBA)-induced rat mammary tumors in vivo (Müllauer et al., manuscript in preparation). Hypothetically, tumor cells may escape the growth restriction which is active on normal tissues and at the same time may prevent growth or even induce death of their healthy neighbour cells.

The recent developments in our knowledge on the network of extrin-sic and intrinsic factors controlling apoptosis are revealing numerous potential targets for therapeutic intervention (for review see Bursch 1994; Dive et al. 1992; Dive and Hickman 1991; Grunicke and Hofmann 1992; Eastman 1990; Hickman 1992; Workman et al. 1992). Thus, receptor and signal transduction pathways might be used for selective induction of apoptosis in target cells. Furthermore, therapeutic modulation of such signal transduction pathways in tumor cells might facilitate the induction of apoptosis by hormones, antibodies, and cytostatic drugs. The genetic control of apoptosis might provide targets for antigene/antisense drugs to block proliferation and to direct cells into apoptosis. Such approaches may eventually help to overcome the selection for hormone-independent or other drug-resistant tumor cells currently confounding tumor therapy.

References

Allan DJ, Howell A, Roberts SA, Williams GT, Watson RJ, Coyne JD, Clarke RB, Laidlaw IJ and Potten CS (1992) Reduction of apoptosis relative to mitosis in histologically normal epithelium accompanies fibrocystic change and carcinoma of the premenopausal human breast. J Pathol 167:25–32

Arends MJ, Morris RG and Wyllie AH (1990) Apoptosis – the role of endonuclease. Am J Pathol 136:593–607

Bardon S, Vignon F, Montcourier P, Rochefort H (1987) Steroid receptor-mediated cytotoxicity of an antiestrogen and antiprogestin in breast cancer cells. Cancer Res 47:1441–1448

Bellomo G, Perotti M, Taddei F, Mirabelli F, Finardi G, Nicotera P, Orrenius S (1992) Tumor necrosis factor alpha induces apoptosis in mammary adenocarcinoma cells by an increase in intracellular free Ca^{2+} concentration and DNA fragmentation. Cancer Res 52:1342–1346

Bissonette R, Echeverri F, Mahboubi A, Green D (1992) apoptotic cell death induced by c-myc is inhibited by bcl-. Nature 359:552–554

Brown DG, Sun XM, Cohen GM (1993) Dexamethasone-induced apoptosis involves cleavage of DNA to large fragments prior to internucleosomal fragmentation. J Biol Chem 268:3037–3039

Bursch W (1994) Apoptosis and cancer therapy. In: Workman P (ed) New approaches in cancer pharmacology: drug design and development, vol II. Springer, Berlin Heidelberg New York (European School of Oncology monographs) (in press)

Bursch W, Lauer B, Timmermann-Trosiener I, Barthel G, Schuppler J, Schulte-Hermann R (1984) Controlled cell death (apoptosis) of normal and

putative preneoplastic cells in rat liver following withdrawal of tumor promoters. Carcinogenesis 5:453–458

Bursch W, Taper HS, Lauer B, Schulte-Hermann R (1985) Quantitative histological and histochemical studies on the occurrence and stages of controlled
cell death (apoptosis) during regression of rat liver hyperplasia. Virch Arch
Cell Pathol 50:153–166

Bursch W, Düsterberg B, Schulte-Hermann R (1986) Growth, regression and
cell death in rat liver as related to tissue levels of the hepatomitogen cyproterone acetate. Arch Toxicol 59:221–227

Bursch W, Paffe S, Putz B, Barthel G, Schulte-Hermann R (1990) Determination of the length of the histological stages of apoptosis in normal liver and
in altered hepatic foci of rats. Carcinogenesis 11:847–853

Bursch W, Liehr JG, Sirbasku D, Putz B, Taper H, Schulte-Hermann R (1991)
Control of cell death (apoptosis) by diethylstilbestrol in an estrogen dependent kidney tumor. Carcinogenesis 12:855–860

Bursch W, Oberhammer F, Schulte-Hermann R (1992) Cell death and its protective role against disease. Trends Pharmacol Sci 13:245–251

Bursch W, Oberhammer F, Jirtle RL, Askari M, Sedivy R, Grasl-Kraupp B,
Purchio AF (1993) Transforming growth factor-β1 as a signal for induction
of cell death by apoptosis. Br J Cancer 67:531–536

Bursch W, Gleeson T, Kleine L, Tenniswood M (1994) Expression of clusterin (testosterone-repressed prostate message-2) mRNA during growth and
regression of rat liver. Arch Toxicol (in press)

Bursch W, Kienzl H, Ellinger A, Schulte-Hermann R Cell death in cultured
human mammary carcinoma cells (MCF-7) after treatment with the antiestrogens tamoxifen and ICI 164 384. (Submitted)

Buttyan R, Zakeri Z, Lockshin RA, Wohlgemuth D (1988) Cascade induction
of c-fos, c-myc, and heat shock 70 k transcripts during regression of the rat
ventral prostate gland. Mol Endocrinol 2:650–657

Clarke AR, Purdie CA, Harrison DJ, Morris RG, Bird CC, Hooper ML, Wyllie
AH (1993) Thymocyte apoptosis induced by p53-dependent and independent pathways. Nature 362:849–852.

Clarke PH (1990) Developmental cell death: morphological diversity and
multiple mechanisms. Anat Embryol 181:195–213

Cohen GM, Sun XM, Snowden RT, Dinsdale D, Skileter DN (1992) Key morphological features of apoptosis may occur in the absence of internucleosomal DNA fragmentation. Biochem J 286:331–334

Cohen JJ, Duke R, Fadok V, Sellins K (1992) Apoptosis and programmed cell
death in immunity. Annu Rev Immunol 10:267–293

Collins RJ, Harmon V, Gobé GC, Kerr JFR (1992) Internucleosomal DNA
cleavage should not be the sole criterion for identifying apoptosis. Int J
Radiat Biol 61:451–453

Columbano A, Ledda-Columbano GM, Rao PM, Rajalakshmi S, Sarma DSR (1984) Occurrence of cell death (apoptosis) in preneoplastic and neoplastic liver cells: a sequential study. Am J Pathol 116:441–446

Columbano A, Ledda-Columbano GM, Coni PP, Faa G, Liguori C, Santa Cruz G, Pani P (1985) Occurrence of cell death (apoptosis) during the involution of liver hyperplasia. Lab Invest 52:670

Dini L, Autuori F, Lentini A, Oliverio S, Piancentini M (1992) The clearance of apoptotic cells in the liver is mediated by the asialoglycoprotein receptor. FEBS Lett 296:174–178

Dive C, Hickman JA (1991) Drug-target interactions: only the first step in the commitment to a programmed cell death. Br J Cancer 64:192–196

Dive C, Evans CA, Whetton AD (1992) Induction of apoptosis – new targets for cancer chemotherapy. Semin Cancer Biol 3:417–427

Eastman A (1990) Activation of programmed cell death by anticancer agents: cisplatin as a model system. Cancer Cells 2:275–280

Ellis RE, Yuan J, Horvitz HR (1991) Mechanisms and functions of cell death. Annu Rev Cell Biol 7:663–698

Evan GI, Wyllie AH, Gilbert CS, Littlewood TD, Land H, Brooks M, Waters CM, Penn LZ, Hancock DC (1992) Induction of apoptosis in fibroblasts by c-myc protein. Cell 69:119–128

Fadok VA, Voelker DR, Campbell PA, Cohen JJ, Bratton DL, Henson PM (1992) Exposure of phosphatidlyserine on the surface of apoptotic bodies triggers specific recognition and removal by macrophages. J Immunol 148:2207–2216

Fanidi A, Harrington EA, Evan GI (1992) Cooperative interaction between c-myc and bcl-2 proto-oncogenes. Nature 359:554–556

Farber E, Cameron R (1980) The sequential analysis of cancer development. Adv Cancer Res 31:125–225

Farber E, Verbin RS, Lieberman M (1972) Cell suicide and cell death. In: Aldridge N (ed) A symposium on mechanisms of toxicology. Macmillan, New York, pp 163–173

Fesus L (1991) Apoptosis fashions T and B cell repertoire. Immunol Lett 30:277–282

Filipski J, Leblanc J, Youdale T, Sikorska M, Walker PR (1990) Periodicity of DNA folding in higher order chromatin structures. EMBO J 9:1319–1327

Fukuda K, Kojiro M, Chiu, JF (1993) Induction of apoptosis by transforming growth factor-$\beta 1$ in the rat hepatoma cell line McA-RH7777: a possible association with tissue transglutaminase expression. Hepatology 18:945–953

Garcea R, Daino L, Pascale R, Simile M, Puddu M, Frassetto S, Cozzolino P, Seddaiu MA, Gaspa L, Feo F (1989) Inhibition of promotion and persistent nodule growth by S-adenosyl-L-methionine in rat liver carcinogenesis: role of remodeling and apoptosis. Cancer Res 49:1850–1856

Glücksmann A (1951) Cell death in normal vertebrate ontogeny. Biol Rev Camb Philos Soc 26:59–86

Grasl-Kraupp B, Huber W, Taper H, Schulte-Hermann R (1991) Increased susceptibility of aged rats to hepatocarcinogenesis by the peroxisome proliferator nafenopin and the possible involvement of altered liver foci occurring spontaneously. Cancer Res 51:666–671

Grasl-Kraupp B, Huber W, Schulte-Hermann R (1993) Are peroxisome proliferators tumour promoters in rat liver? In: Gibson CG, Lake B (eds) Monograph on peroxisome proliferation. Taylor and Francis, London, pp 667–693

Grasl-Kraupp B, Bursch W, Ruttkay-Nedecky B, Wagner A, Lauer B, Schulte-Hermann R (1994) Food restriction eliminates preneoplastic cells through apoptosis and antagonizes carcinogenesis in rat liver. PNAS 91:9995–9999

Grunicke H, Hofmann J (1992) Cytotoxic and cytostatic effects of antitumor agents induced at the plasma membrane level. Pharmacol Ther 55:1–30

Gullino PM (1980) The regression process in hormone-dependent mammary carcinomas. In: Iacobelli S, King RBJ, Lindner HR, Lippman ME (eds) Hormones and cancer. Raven, New York, pp 271–279

Hengartner MO, Ellis RE, Horvitz HR (1992) Caenorhabditis elegans gene ced-9 protects cells from programmed cell death. Nature 356:494–499

Hickman JA (1992) Membrane and signal transduction targets. In: Workman P, D'Incalci M (eds) New approaches in cancer pharmacology: drug design and development, vol I. Springer, Berlin Heidelberg New York, pp 33–46 (European School of Oncology monographs)

Hockenberry D, Nunez G, Milliman C, Schreiber RD, Korsmeyer SJ (1990) Bcl-2 is an inner mitochondrial membrane protein that blocks programmed cell death. Nature 348:334–336, 1990.

Isaacs JT (1984) Antagonistic effect of androgen on prostatic cell death. Prostate 5:545–557

Jenne D, Tschopp J (1992) Clusterin: the intriguing guises of a widely expressed glycoprotein. TIBS 17:154–159

Kerr JFR, Wyllie AH, Currie AR (1972) Apoptosis: a basic biological phenomenon with wide-ranging implications in tissue kinetics. Br J Cancer 26:239–257

Kirszbaum L, Sharpe JA, Murphy B, d'Apice AJF, Classon B, Hudson P, Walker ID (1989) Molecular cloning and characterization of the novel, human complement-associated protein, SP-40,40: a link between the complement and reproductive systems. EMBO J 8:711–718

Korsmeyer SJ (1992) Chromosomal translocation in lymphoid malignancies reveal novel proto-oncogenes. Annu Rev Immunol 10:785–807

Kyprianou N, English HF, Isaacs JT (1990) Programmed cell death during regression of PC-82 human prostate cancer following androgen ablation. Cancer Res 50:3748–3753

Kyprianou N, English HF, Davidson NE, Isaacs JT (1991) Programmed cell death during regression of the MCF-7 human breast cancer following estrogen ablation. Cancer Res 51:162–166

Lanzerotti LH, Gullino PM (1972) Activity and quantity of lysosomal enzymes during mammary tumor regression. Cancer Res 32:2679–2685

Laster SM, Wood JG, Gooding LR (1988) Tumor necrosis factor can induce both apoptotic and necrotic forms of cell lysis. J Immunol 141:2629–2633

Léger J, Le Guellec R, Tenniswood PR (1988) Treatment with antiandrogens induces an androgen repressed gene in the rat ventral prostate. Prostate 13:131–142

Levi-Montalcini R (1987) The nerve growth factor thirty-five years later. In Vitro Cell Dev Biol 23:227–283

Lin JK, Chou CK (1992) In vitro apoptosis in the human hepatoma cell line induced by transforming growth factor beta 1. Cancer Res 52:385–388

Lockshin RA, Beaulaton J (1974) Programmed cell death. Cytochemical evidence for lysosomes during the normal breakdown of the intersegmental muscles. J Ultrastruct Res 46:43–62

Lockshin RA, Williams CM (1964) Programmed cell death. II. Endocrine potentiation of the breakdown of the inter-segmental muscles of silk moths. J Insect Physiol 10:643–649

Lockshin RA, Williams CM (1965) Programmed cell death. I. Cytology of degeneration in the intersegmental muscles of the Pernyi silk moth. J Insect Physiol 11:123–133

Lowe SW, Schmitt EM, Smith SW, Osborne BA, Jacks T (1993) p53 is required for radiation-induced apoptosis in mouse thymocytes. Nature 362:847–849

Luebeck EG, Moolgavkar SH, Buchmann A, Schwarz M (1991) Effects of polychlorinated biphenyls in rat liver: quantitative analysis of enzyme-altered foci. Toxicol Appl Pharmacol 111:469–484

MacDonald HR, Lees RK (1990) Programmed cell death of autoreactive thymocytes. Nature 343:642–644

Martz E, Howell DM (1989) CTL: virus control cells first and cytoloytic cells second? DNA fragmentation, apoptosis and the prelytic halt hypothesis. Immunol Today 10:79–86

Michna H, Nishino Y, Neef G, McGuire WL, Schneider MR (1992) Progesterone antagonists: tumor-inhibiting potential and mechanism of action. J Steroid Biochem Mol Biol 41:339–348

Montpetit ML, Lawless KR, Tenniswood (1986) Androgen-repressed messages in the rat ventral prostate. Prostate 8:25–36

Moolgavkar SH, Luebeck EG, De Gunst M, Port RE, Schwarz M (1990) Quantitative analysis of enzyme-altered foci in rat hepatocarcinogenesis experiments I: single agent regimen. Carcinogenesis 11:1271–1278

Oberhammer F, Pavelka M, Sharma S, Tiefenbacher R, Purchio TA, Bursch W, Schulte-Hermann R (1992) Induction of apoptosis in cultured hepatocytes and in regressing liver by transforming growth factor-β1. Proc Natl Acad Sci USA 89:5408–5412

Oberhammer F, Bursch W, Tiefenbacher R, Fröschl G, Pavelka M, Purchio T, Schulte-Hermann R (1993a) Apoptosis is induced by transforming growth factor-β1 within 5 hours in regressing liver without significant fragmentation of the DNA. Hepatology 18:1238–1246

Oberhammer F, Wilson JW, Dive C, Morris ID, Hickman JA, Wakeling AE, Walker PR, Sikorska M (1993b) Apoptotic death in epithelial cells: cleavage of DNA to 300 and/or 50 kb fragments prior to or in the absence of internucleosomal fragmentation. EMBO J 12:3679–3684

Oberhammer F, Fritsch G, Schmied M, Pavelka M, Printz D, Purchio T, Lassmann H (1993c) Condensation of the chromatin at the membrane of an apoptotic nucleus is not associated with activation of an endonuclease. J Cell Sci 104:317–326

Ogasawara J, Watanabe-Fukunaga R, Adachi M, Matsuzawa A, Kasugai T, Kitamura Y, Itoh N, Suda T, Nagata S (1993) Lethal effect of the anti-Fas antibody in mice. Nature 364:806–809

Oltvai ZN, Milliman CL, Korsmeyer SJ (1993) Bcl-2 heterodimerizes in vivo with conserved homolog, Bax, that accelerates programmed cell death. Cell 74(4):609–619

Pascale RM, Marras V, Simile MM, Daino L, Inna G, Bennati S, Carta M, Seddaiu MA, Massarelli G, Feo F (1992) Chemoprevention of rat liver carcinogenesis by S-adenosyl-L-methionine: a long term study. Cancer Res 52:4979–4986

Pitot HC, Sirica AE (1980) The stages of initiation and promotion in hepatocarcinogenesis. Biophys Acta 605:191–215

Popper H, Keppler D (1986) Networks of interacting mechanisms of hepatocellular degeneration and death. In: Popper H, Schaffner F (eds) Progress in liver disease, vol VIII. Grune and Stratton, Orlando, pp 209–236

Redding TW, Schally AV, Radulovic S, Milovanovic S, Szepeshazi K, Isaacs JT (1992) Sustained release formulations of luteinizing hormone-releasing hormone antagonist SB-75 inhibit proliferation and enhance apoptotic cell death of human prostate carcinoma (PC-82) in male nude mice. Cancer Res 52:2538–2544

Rotello RJ, Liebermann RC, Purchio A, Gerschenson LE (1991) Coordinated regulation of apoptosis and cell proliferation by transforming growth factor β1 in cultured uterine epithelial cells. Proc Natl Acad Sci USA 88:3412–3415

Sarraf CE, Bowen ID (1986) Kinetic studies on a murine sarcoma and an analysis of apoptosis. Br J Cancer 54:989–998

Sarraf CE, Bowen ID (1988) Proportions of mitotic and apoptotic cells in a range of untreated experimental tumors. Cell Tissue Kinet 21:45–49

Saunders JW (1966) Death in embryonic systems. Death of cells is the usual accompaniment of embryonic growth and differentiation. Science 54:604–612

Savill J, Dransfield I, Hogg N, Haslett C (1990) Vitronectin receptor-mediated phagocytosis of cells undergoing apoptosis. Nature 343:170–173

Schulte-Hermann R (1977) Two stage control of cell proliferation induced in rat liver by α-hexachlorocyclohexane. Cancer Res 37:166–171

Schulte-Hermann R (1985) Tumor promotion in the liver. Arch Toxikol 57:147–215

Schulte-Hermann R, Schuppler I, Timmermann-Trosiener I, Berger H (1983) The role of growth of normal and preneoplastic cell populations for tumor promotion in rat liver. Environ Health Perspect 50:185–194

Schulte-Hermann R, Bursch W, Fesus L, Kraupp B (1988) Cell death by apoptosis in normal, preneoplastic and neoplastic tissue. In: Feo F, Pani P, Columbano A, Garcea R (eds) Chemical carcinogenesis: models and mechanisms. Plenum, New York, pp 263–274

Schulte-Hermann R, Timmermann-Trosiener I, Barthel G, Bursch W (1990) DNA synthesis, apoptosis and phenotypic expression as determinants of growth of altered foci in rat liver during phenobarbital promotion. Cancer Res 50:5127–5135

Schwall RH, Robbins K, Jardieu P, Chang L, Lai C, Terrell TG (1993) Activin induces cell death in hepatocytes in vivo and in vitro. Hepatology 18:347–356

Schweichel JU, Merker HJ (1973) The morphology of various types of cell death in prenatal tissues. Teratology 7:253–266

Shinagawa T, Yoshioka K, Kamuku S, Wakita T, Ishikawa T, Itoh Y, Takayanagi M (1991) Apoptosis in cultured rat hepatocytes: the effects of tumor necrosis factor alpha and interferon gamma. J Pathol 165:247–53

Szende B, Zalatnai A, Schally AV (1989) Programmed cell death (apoptosis) in pancreatic cancers of hamsters after treatment with analogs of both luteinizing hormone-releasing hormone and somatostatin. Proc Natl Acad Sci USA 83:1643–1647

Szende B, Srkalovic G, Groot K, Lapis K, Schally AV (1990) Regression of nitrosamine-induced pancreatic cancers in hamsters treated with luteinizing hormone-releasing hormone antagonists or agonists. Cancer Res 50:3716–3721

Szepeshazi K, Lapis K, Schally AV (1991) Effect of combination treatment with analogs of luteinizing hormone-releasing hormone (LH-RH) or somatostatin and 5-fluorouracil on pancreatic cancer in hamsters. Int J Cancer 49:260–226

Tenniswood M, Guenette RE, Lakins JL, Mooibroek M, Wong P, Welsh J (1992) Active cell death in hormone dependent tissues. Cancer Metastasis Rev 11:197–220

Thompson AM, Kerr DJ, Steel CM (1991) Transforming growth factor β1 is implicated in the failure of tamoxifen therapy in human breast cancer. Br J Cancer 63:609–614

Valverius EM, Walker-Jones D, Bates SE, Stampfer MR, Clark R, McCormick F, Dickson RB, Lippman ME (1989) Production of and responsiveness to transforming growth factor-β in normal and oncogene-transformed human mammary epithelial cells. Cancer Res 49:6269–6274

Vaux DL, Cory S, Adams JM (1988) Bcl-2 gene promotes haemopoietic cell survival and cooperates with c-myc to immortalize pre-B cells. Nature 335:440–442

Walker PR, Kokileva L, LeBlanc J, Sikorska M (1993) Detection of the initial stages of DNA fragmentation in Apoptosis. Biotechniques 15(6):1032–1040

Wärri AM, Huovinen RL, Laine AM, Marikainen PM, Härkönen PL (1993) Apoptosis in toremifene-induced growth inhibition of human breast cancer cells in vivo and in vitro. J Natl Cancer Inst 85:1412–1418

Workman P, D'Incalci M, Berdel WE, Egorin MJ, Helene C, Hickman JA, Jarman M, Schwartsman G, Sikora K (1992) New approaches in cancer pharmacology: drug design and development. Eur J Cancer 28A:1190–1200

Wyllie AH (1980) Glucocorticoid-induced thymocyte apoptosis is associated with endogenous activation. Nature 284:555–556

Wyllie AH (1985) The biology of cell death in tumors. Anticancer Res 5:131–136

Wyllie AH, Kerr J, Currie A (1980) Cell death: the significance of apoptosis. Int Rev Cytol 68:251–306

Wyllie AH, Morris RG, Smith AL, Dunlop D (1984) Hormone-induced cell death. 2. Surface changes in thymocytes undergoing apoptosis. Am J Pathol 115:426–436

Yanagihara K, Tsumuraya M (1992) Transforming growth factor beta 1 induces apoptotic cell death in cultured human gastric carcinoma cells. Cancer Res 52:4042–4045

Yonish-Rouach E, Resnitzky D, Lotem J, Sachs L, Kimchi A, Oren M (1991) Wild-type p53 induces apoptosis of myeloiod leukemia cells that is inhibited by interleukin-6. Nature 352:345–347

Zakeri ZF, Quaglino D, Latham T, Lockshin RA (1993) Delayed internucleosomal DNA fragmentation in programmed cell death. FASEB J 7:470–478

Zugmaier G, Paik S, Wilding G, Knabbe C, Bano M, Lupu R, Deschauer B, Simpson S, Dickson R, Lippman M (1991) Transforming growth factor β1 induces cachexia and systemic fibrosis without an antitumor effect. Cancer Res 51:3590–3594

6 Modulation of Apoptosis by Oncogenes

P. R. Walker, L. Testolin, U. Armato, N. Marceau, H. Gourdeau, and M. Sikorska

6.1 Introduction

Apoptosis is a physiological pathway by which tissues eliminate unwanted cells. As far as we know every cell possesses the capability to undergo this form of cellular suicide, but since only a few cells actually do undergo apoptosis the pathway must be tightly regulated. There is currently tremendous interest in being able to understand the nature of these controls so that the cell death pathway can be modulated for therapeutic purposes. Thus, increasing the rate of the endogenous cell death pathway would be the preferred way to remove cancer cells whereas downregulation of apoptosis may be beneficial in the treatment

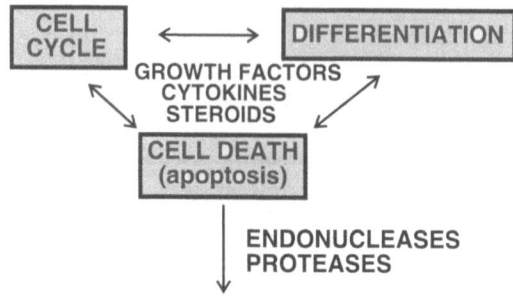

Destruction of nucleus

Fig. 1. Apoptosis is a cellular state similar to cell cycle and differentiation

of a number of diseases, including AIDS and neurodegenerative disorders.

To date cells have been considered to exist in two major states; they are either in the cell cycle or in G_0 where they undergo differentiation. Tissue-specific growth and differentiation factors control the movement of cells into and out of these two states. Progression through, or existence in, each state is controlled by the many intracellular regulatory molecules that effect the changes in gene expression required to maintain that state. Our views on the molecular mechanisms of neoplasia are, to a large extent, based upon this view of cellular states. Thus cells are considered to acquire mutations in certain critical genes that either leave cells inappropriately in the cell cycle or force differentiated cells to dedifferentiate and re-enter the cell cycle.

Apoptosis must now be considered a third state into which cells can enter or be rescued from (Fig. 1). It is now clear that cells are able to exit the cell cycle and undergo apoptosis or to enter the cell death pathway directly from the differentiated state. The movement of cells from these states into apoptosis must inherently be tightly controlled and a number of hormones and cytokines that repress cell death have been identified as well as factors such as tumour necrosis factor and transforming growth factor β which are potent inducers of cell death. Since a decrease in the rate of cell death of a population can have just as equal an impact on its rate of growth as increased proliferation, it is clear that our views on tumorigenicity must be revised (Williams 1991). Mutations in gene products involved in the regulation of apoptosis can trigger an imbalance in tissue turnover, leading to the formation of a cancerous growth. A great deal is known about the regulation of the cell cycle and the factors responsible for maintaining cells in various differentiated states are also being elucidated. In general, cells respond to cell-type or tissue-specific hormone, cytokine or growth factor signals that either drive them into the cell cycle or maintain them in their differentiated state. Intracellular signal transduction pathways transmit this information to the nucleus so that the appropriate genes can be expressed (or repressed). It is now evident that the apoptotic pathway is under genetic control (see Gourdeau and Walker 1994 and references therein) and also requires gene expression in most instances. A great deal of effort is being spent on trying to identify unique gene products that are expressed in this cellular programme. In addition the signal transduction pathways responsible for the control of these genes are being studied.

Our understanding of the regulation of the cell cycle has been greatly impacted by the identification of the genes of various viruses that are responsible for oncogenic transformation. In general, viral oncogenes are mutated forms of endogenous genes (proto-oncogenes) involved in the normal control of cell cycle progression. The inappropriate expression of one or more of these genes, for example *myc, src, ras* or *fos*, can lead to neoplastic transformation by forcing cells to remain in the cell cycle. Studies on these oncogenes and their normal cellular homologues has allowed us to delineate, in many cases, the sequence of events from the initial binding of growth factor to its plasma membrane receptor through to the activation of transcription factors that control specific patterns of gene expression in the nucleus. Although the equivalent regulators of apoptosis have not yet been characterised, the recent identification of the *bcl-2* oncogene (Hockenbury et al. 1991) as a repressor of cell death, as opposed to an upregulator of proliferation, indicates that apoptosis-specific regulatory gene products do exist and that their altered expression can result in tumour formation. For example, the t(14:18) translocation of the *bcl-2* gene results in B cell lymphoma and *bcl-2* expression is linked to the emergence of hormone-independent prostate cancer (McDonnell et al. 1992,1993). Moreover, the roles of proto-oncogenes such as p53 or c-*myc*, whose function in cell cycle progression was never well defined, are becoming better understood, and they appear to play some role in determining the susceptibility of cells to apoptosis. The latter examples suggest an intimate relationship between cell cycle regulators and the regulators of the pathway of apoptosis (Amati et al. 1993; Lowe et al. 1994). Less clear are the relationships between regulators of differentiated function and apoptosis and on the relationship between agents causing neoplastic transformation and apoptosis.

Overexpression of the *bcl*-2 oncogene has been shown to protect a number of different types of cells from apoptosis. These include the protection of human BAF-3 cells from withdrawal of interleukin 3 (IL-3; Oliver et al. 1993), L929 cells against tumour necrosis factor (Hennet et al. 1993) and neurotrophic factor-dependent neurons against factor withdrawal (Allsop et al. 1993). *Bcl*-2 overexpression also renders the BAF3 cells resistant to 5-fluorouracil (5FU) and etoposide (Oliver et al. 1993; Ascaso et al. 1994). Similarly, various lymphoma cells were protected against the cytotoxic effects of 5FU and other inhibitors of thymidylate synthetase (Fisher et al. 1993). In the latter study the "classical" drug resistance mechanisms were ruled out, confirming that *bcl*-2 acts in a novel fashion by intefering with the activation of apoptosis. However, *bcl*-2 does not suppress apoptosis in all circumstances. Most notable, with regard to this study, is the failure of *bcl*-2 to protect cytotoxic T cell line (CTLL) cells against IL-2 withdrawal.

The discovery that chemotherapeutic drugs, many of which were designed as, or considered to be, inhibitors of cell cycle progression, were also powerful inducers of apoptosis (Walker et al. 1991) has led to a reconsideration of the way in which cancers can be controlled. Two general strategies may be contemplated, the first being to design drugs that directly target molecules involved in the induction of apoptosis and the second to remove the trophic factors which normally repress cell death. In the latter case removal of trophic factor, or interference with its signals, may result in cells undergoing apoptosis or at least becoming more sensitive to undergoing apoptosis following drug treatment. Quite clearly, if a drug is designed to remove cells by triggering apoptosis the presence of hormones or trophic agents that may repress the ability of the drug to activate the cell death pathway would decrease the therapeutic potential of the drug. An understanding of the control of apoptosis is essential in order to identify new molecular targets for chemotherapy and to understand how cancerous cells can be eliminated by specifically depleting them of their survival factors. We have been trying to answer some of these questions by studying the pathways by which chemotherapeutic agents, particularly those inflicting DNA damage, induce apoptosis (Walker et al. 1991; Roy et al. 1992). In addition, we have been examining how tissues control apoptosis and the extent to which alterations in the rate of apoptosis contribute to the transformed phenotype and affect the efficacy of chemotherapeutic drugs (Walker et al. 1993a,b).

6.2 Apoptosis in Oncogene-Transfected Rat Liver Cells

To specifically study the effects of overexpression of various oncogenes on apoptosis we have been using early passage T51B rat liver epithelial cells that were stably transfected with >mycmyc, >rasras, myc + ras, middle T (MT) antigen of polyoma virus or MT + myc (Royal et al. 1992). Growth in soft agar and tumour formation in syngeneic Fisher rats showed that neither myc alone nor ras alone transformed the cells, but the combination of myc + ras led to both growth in soft agar and tumour formation in vivo. The cells were also efficiently transformed with MT and the combination of MT + myc was also tumorigenic. This panel of oncogene-transfected cells is useful for determining the effects of overexpression of the oncogenes on the susceptibility of cells to apoptosis as well as establishing the role that apoptosis may play in determining the transformed phenotype. These studies are particularly relevant to hormone-dependent cancers and more specifically to breast cancer since MT has been shown to induce aggressive adenocarcinomas of the breast in transgenic animals expressing MT (Guy et al. 1994).

Transfection of the T51B cells with either myc alone or MT alone produced dramatic, but contrasting, effects on the growth rate of the cell population. Cells transfected with myc grew much more slowly than the parental cell line that had been transfected with vector only (T51B-neo). In contrast, cells transfected with MT proliferated more rapidly than the parental cell, reaching much higher cell densities in culture (Fig. 2a) and producing many large colonies in soft agar. The presence of myc in the MT + myc co-transfected cells dramatically reduced the rate of cell growth back to that of myc alone, but the cells still produced colonies in soft agar although they were much fewer and smaller. The rate of

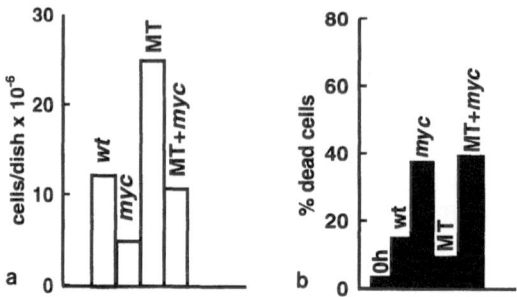

Fig. 2a,b. Rates of proliferation (**a**) and apoptosis (**b**) in transfected T51B liver cells compared to control cells (*Oh*). In **a**, cells were seeded at equal density and counted 72 h later. In **b**, serum was withdrawn and the percentage of dead cells was measured after 24 h. *MT*, middle T antigen; *wt*, untransfected wild-type cells

proliferation of cells transformed with *ras* or *myc + ras* were not re-markably different from those of the parental cell line.

We next evaluated the sensitivity of the transfected cells to apoptosis following serum withdrawal (Fig. 2b). Once again, there were dramatic differences in the rates at which the cells underwent apoptosis. The cells transfected with *myc*, which showed a slower rate of proliferation in the presence of serum, underwent apoptosis rapidly when serum was with-drawn. In contrast, the MT-transfected cells, which had a high rate of cell proliferation in the presence of serum, were resistant to undergoing apoptosis when serum was withdrawn. The simultaneous presence of *myc* in the MT + *myc* co-transfected cells caused the rate of cell death to dramatically increase. The *ras*-transformed cells showed a rate of apop-tosis similar to that of the parental cell line. Thus, there was an inverse relationship between the rate of growth of the population and the rate at which the cells underwent cell death. A similar relationship was found in cells transfected with mutated (T24) H-*ras* or c-*myc* (Arends et al. 1993). Most notably in these studies the MT antigen of polyoma virus was particularly effective at downregulating apoptosis.

6.3 Molecular Basis of the Altered Rates of Apoptosis

A similar phenomenon of *myc*-induced sensitivity to apoptosis has been observed in other cells (Arends et al. 1993; Bennett et al. 1994). These cells grew at different rates in the presence of serum and rapidly underwent apoptosis when serum levels were reduced or growth factor removed. In the case of the vascular smooth muscle cells (Bennett et al. 1994) they failed to leave the cell cycle upon serum withdrawal and continued to proliferate until they underwent apoptosis. The mechanism by which *myc* increases susceptibiliy to apoptosis is not clear. Although *myc* overexpression in the T51B liver cells dramatically reduced growth and increased the rate of apoptosis, it did not affect neoplastic transformation. Thus, *myc* is able to contribute to cellular transformation along with *ras*, but the transformed cell population acquires an increased susceptibility to undergoing apoptosis. This is a particularly dramatic example of a cell population that can proliferate and form a tumour even though the rate of apoptosis is also dramatically increased. It appears that in the presence of continued *myc* expression cells cannot leave the cell cycle (*myc* is normally downregulated in cells as they become quiescent), leaving them in an unstable position. This instability will lead them to undergo apoptosis directly from the cell cycle as growth factors are depleted. In this regard the cells are similar to cytotoxic T cells which cannot re-enter G_0 upon withdrawal of IL-2 (see below).

Transfection of the T51B cells with the MT antigen led to a dramatic increase in cell growth which seemed to be due in large part to the repression of apoptosis. Thus in the presence of serum the cells were able to reach a high cell density by repressing the apoptosis that is normally seen in the parental cell line as it reaches confluence. Therefore, MT can simultaneously transform the cells and render them resistant to apoptosis, producing a particularly rapidly growing cell population. The same characteristic high rate of cell growth is seen in the multifocal adenocarcinomas found in the breast tissue of mice carrying the MT gene under the control of the MMTV promoter (Guy et al. 1994). However, sensitivity to cell death is restored in cells co-transfected with the *myc* oncogene. It is unlikely, therefore, that MT is transforming cells merely by repressing apoptosis in a *bcl*-2-like fashion (see below) since when the rate of apoptosis is dramatically increased by co-transfection with *myc* the cell population remains tumorigenic. The fact that far fewer colonies are formed in soft agar suggests that many cells, perhaps those with a high *myc*:MT expression ratio, die before forming a colony and that colonies are only formed in those cells with a rate of apoptosis that does not exceed the rate of MT-stimulated proliferation.

To assess whether the markedly different rates of cell death seen in these transfected cells were due to the differential expression of the *bcl*-2 oncogene we measured the expression of this gene along with the related *bcl*-x gene and the *bax* gene, which codes for a *bcl*-2 binding protein (Oitavi et al. 1993). In agreement with our other studies on liver cells (Gourdeau and Walker 1994) the *bcl*-2 gene was not expressed in the parental T51B rat liver cell line or in any of the transfected cells. Therefore, differences in susceptibility to apoptosis could not be accounted for by the expression of *bcl*-2. The parental cells and all the transfected cells did, however, express high constitutive levels of the *bax* gene although there was no correlation between mRNA levels and susceptibility to apoptosis. This observation ruled out the possibility that MT was suppressing apoptosis by downregulating bax All of the cells with the exception of the MT-transfected cells expressed low levels of *bcl*-x. The *bcl*-x gene was expressed in MT + *myc* co-transfected cells. Therefore, the cells that were resistant to apoptosis did not express *bcl*-x. The role of the *bcl*-x gene in the control of apoptosis in these cells is presently being studied more extensively. It is clear, therefore, that in liver cells susceptibility to apoptosis is controlled by a *bcl*-2-independent mechanism. Furthermore, the MT antigen is able to activate this independent process and completely repress apoptosis in liver cells.

6.4 Mechanism of Repression of Apoptosis in MT-Transfected Cells

Although such a dramatic effect of MT on apoptosis has not been observed before there are some clues on the mechanism by which it might be acting, based upon its known mechanism of action in relation to our previous studies on repression of apoptosis in growth factor-dependent cells (Walker et al. 1993a). MT antigen is known to bind to intracellular tyrosine kinases of the *src* family (reviewed in Whitfield 1990). These include *src* itself and the closely related growth factor-associated kinases *fyn* and *yes*. This family of tyrosine kinases plays an essential role in growth factor signal transduction in a number of cell types. c-*src* is not directly attached to the intracellular domain of any particular receptor, but being myristylated, it is anchored in the plasma membrane adjacent to a number of growth factor receptors. Binding of MT to *src, fyn or yes* chronically activates these kinases and simulates the continuous presence of growth factor. Such cells are then perpetually in the cell cycle. The MT antigen also binds to and activates phosphatidylinositol kinase, leading to the production of inositol 1,4,5-triphosphate (IP_3), but the role, if any, that this plays in transformation is not clear. It has been established in the mouse transgene model (Guy et al. 1994) that *src* is required for the formation of breast adenocarcinomas. Interestingly, it may not be necessary for the formation of hemangiomas in similar transgenes (Thomas et al. 1993). It is possible in the latter cells that either *fyn* or *yes* can substitute for *src* deficiency. The effects of chronic activation of pp60-*src* have been studied in many cell types. One consequence of this is the activation of both the expression and the redistribution of protein kinase C (PKC) to the plasma membrane (reviewed in Whitfield 1990). This is particularly relevant to the repression of apoptosis since, as described below, we have established a role for both PKC and tyrosine kinase activation in the control of apoptosis and cell cycle progression in CTLL cells.

6.5 Role of Tyrosine Kinases and PKC in the Hormonal Control of Apoptosis in Lymphocytes

Studies on the mechanisms by which cytokines, primarily IL-2, repress apoptosis and stimulate cell cycle progression in lymphocytes (Walker et al. 1993a) have revealed several similarities to the proposed mechanism of action of MT. Cytotoxic T cells undergo clonal expansion following engagement of their T cell receptor and become dependent upon IL-2 for growth and survival. As IL-2 levels subside the activated T cells undergo apoptosis so that unwanted lymphocytes are removed. Thus, although the peripheral lymphocytes are in a quiescent G_0 phase before activation they cannot re-enter G_0 as IL-2 levels decline and undergo apoptosis. In this regard they are similar to transformed cells and are an excellent model system for studying the regulation of apop-

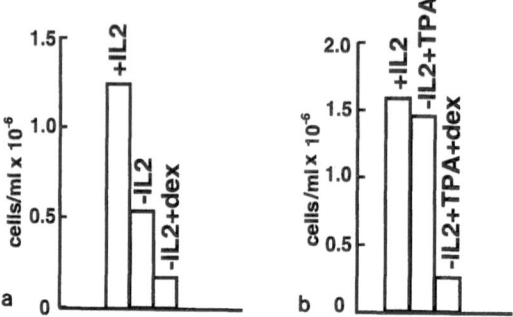

Fig.3a,b. Apoptosis in CTLL cells. **a** Cell growth or apoptosis in the presence of interleukin-2 (*IL-2*), absence of IL-2 and –IL-2 + dexamethasone. In **b**, either tetradecanoylphorbol 13-acetate (*TPA*) or TPA + dexamethasone (*dex*) were added at the time of IL-2 withdrawal and the change in cell number measured 48 h later

tosis and the relationship between the cell cycle and cell death. An established cytotoxic T cell line, the IL-2-dependent CTLL cell, mimics this situation in vitro (Gillis and Smith 1977). This cell line requires the continuous presence of IL-2 to progress through G_1 and to pass the $G_1\backslash S$ restriction point and initiate DNA replication. If, following mitosis, the level of IL-2 is insufficient to support a further round of cell division the cells accumulate at the beginning of G_1. After several hours without IL-2 the cells become irreversibly committed to undergo apoptosis and more than half of the cells have undergone apoptosis by 48 h (Fig. 3a). These cells have an added advantage for cell death studies since glucocorticoids can also induce apoptosis. For most cell types, glucocorticoids stimulate anabolic pathways of metabolism, but in thymocytes and lymphocytes they induce apoptosis. Thus, in the absence of IL-2, glucocorticoids greatly increase the rate at which the cells undergo apoptosis with virtually the entire population being dead in less than 24 h (Fig. 3a). However, this only occurs in the absence of IL-2. In the presence of the cytokine the cells are resistant to the effects of glucocorticoids. Thus, IL-2 not only protects against the apoptosis that occurs as a result of a failure to continue cell cycle progression, but also makes them resistant to the effect of glucocorticoids, suggesting that multiple signals are likely to be involved.

IL-2 is required, therefore, to generate all the signals necessary to repress apoptosis, promote cell cycle progression and confer resistance to glucocorticoids. All these signals are generated following the interaction of the cytokine with its surface receptor. The IL-2 receptor (IL-2R) consists of two subunits of p55 (IL-2R α) and p75 (IL-2R β) and is unique amongst growth factor receptors in that it has only a very short intracellular domain (13 residues for the α subunit and 286 for the β subunit) and no enzymatic activity has been ascribed to this region (Smith 1989). Most other growth factor receptors have an intrinsic tyrosine kinase activity associated with the intracellular domain of the protein. However, two signal transduction pathways have been found to respond to the interaction between IL-2 and IL-2R. The first (which is still controversial in relation to proliferation) is the PKC pathway and the second is the activation of tyrosine kinase(s).

Evidence for a functional role for PKC in CTLL cells initially came from studies which showed that the phorbol ester tetradecanoylphorbol 13-acetate (TPA) maintained the viability of cells deprived of IL-2 for up to 10 h (Rodriguez-Tarduchy and Lopez-Rivas 1989). Thus PKC activation prevented the cells from moving into apoptosis, suggesting that IL-2 may switch cell death off by a PKC-mediated phosphorylation step. Similarly, phorbol esters have been shown to inhibit ionophore and glucocorticoid-induced DNA fragmentation in immature rat thymocyte primary cultures (McConkey et al. 1989). Similar studies on other cells, for example confluent C3H-10T1/2 cells which undergo apoptosis when incubated for up to 72 h without serum, have produced similar results. A population of the latter cells became nonadherent and extensive DNA degradation typical of apoptosis was found in the released cells (Kanter et al. 1984; Tomei et al. 1988). Epidermal growth factor or TPA was able to repress cell death and to induce proliferation. Significantly, in these cells TPA also conferred resistance to apoptosis that normally occurs after ionising radiation. Similarly, endothelial fibroblasts, which like many cells also undergo apoptosis in the absence of serum (Araki et al. 1990a,b), can be rescued by fibroblast growth factor or TPA. It is clear from these studies that PKC activation can replace several growth factors (IL-2, epidermal growth factor and fibroblast growth factor) and inhibit the active cell death pathway, suggesting that it plays a fundamental role in the switching process between viability and cell death.

We have extended these studies and found that TPA can protect CTLL cells from undergoing apoptosis for at least 48 h after IL-2 withdrawal (Fig. 3b). However, the cells did not proliferate, indicating that the activation of PKC by TPA – although it is sufficient to prevent apoptosis – cannot induce cell cycle progression. Thus, the cells remain arrested in early G_1 in what might be considered a "pseudo" G_0 state. Interestingly, phorbol esters cannot prevent glucocorticoid-induced cell death and 80% of the cells were dead by 24–48 h in the presence of TPA and dexamethasone (Fig. 3b). IL-2 must, therefore, generate additional signals which (a) allow the cells to undergo cell cycle progression and (b) cause them to become resistant to glucocorticoids.

In human T cells tyrosine phosphorylation of a number of proteins is essential for IL-2 to drive proliferation since tyrphostin, a tyrosine kinase inhibitor, prevents cell cycle progression (Munoz et al. 1991). The major cytoplasmic protein phosphorylated on tyrosine residues was a protein considered to be the 42-kDa microtubule-associated protein 2 kinase (MAP kinase), suggesting that MAP kinases may play a major downstream role in IL-2 signal transduction. However, additional substrates are phosphorylated. Tyrosine kinase(s), therefore, appear to be the other principal transducers of information from the IL-2R.

We have studied the effects of tyrphostins, which are specific inhibitors of tyrosine kinases, on apoptosis in CTLL cells (Walker et al. 1993a). When tyrphostin was added along with IL-2 there was no further increase in cell number, showing that cell cycle progression was arrested. However, there was only a modest increase in the number of apoptotic cells. Thus, inhibition of tyrosine kinases resulted in G_1 arrest, but the cells did not immediately undergo apoptosis. Indeed, when tyrosine kinases are inhibited cell cycle progression is blocked, but the cells do not immediately undergo apoptosis unless PKC is also inhibited. Thus, incubation of cells in the simultaneous presence of IL-2 and a specific membrane-localised myristoylated peptide inhibitor of PKC causes the population to rapidly undergo cell death confirming that PKC activity is required to disable the cell death pathway. Interestingly, cells incubated in the presence of tyrphostin reacquired sensitivity to glucocorticoids even when IL-2 remained in the medium. Thus, the ability of the cells to undergo cell cycle progression and to be resistant to the effects of glucocorticoids is conferred by tyrosine kinase-mediated signals.

6.6 Role of the *fos* and *jun* Oncogenes and AP-1 in Repression of Apoptosis in Lymphocytes

Phorbol esters activate PKC and transduce signals to the nucleus to effect changes in gene expressions through a common DNA regulatory element in their promoters, the TPA-responsive element (TRE), which is the binding site for the transcription factor AP-1 (Angel et al. 1987). The AP-1 transcription factor complex is composed primarily of the gene products of the proto-oncogenes *c-fos* and *c-jun*, consisiting mainly of *Jun-Jun* homodimers or *Jun-Fos* heterodimers, although other members of the *Fos* and *Jun* family of transcription factors also form complexes (Ryseck and Bravo 1991). Therefore, although collectively referred to as AP-1, several complexes may be involved in mediating changes in gene expression at the TRE. Since the AP-1 transcription factor complex is required for cells to undergo G_1 progression (Riabowol et al. 1992), it is likely to play a key role in switching between apoptosis and the cell cycle. The recent findings (Diamond et al. 1990; Miner et al. 1991; Ponta et al. 1992) that AP-1 can also modify the expression of glucocorticoid-inducible genes (and vice versa) through a complex series of interactions with the glucocorticoid receptor (GR) and the cognate DNA binding elements (TRE and GRE) are also relevant to studies on glucocorticoid-induced apoptosis. Because of the relationships between AP-1 and cell cycle progression as well as AP-1 and the GR we examined the role of AP-1 in switching between the cell cycle and apoptosis in CTLL cells in the presence and absence of IL-2 and glucocorticoids. In the presence of IL-2 the cells have high levels of AP-1 DNA-binding activity which is compatible with its postulated essential role in cell cycle progression. Withdrawal of IL-2 led to a decline in AP-1 DNA-binding activity which preceded DNA fragmentation (Walker et al. 1993a). In these cells the glucocorticoid analogue dexamethasone dramatically increased the rate of loss of AP-1 DNA-binding activity in IL-2-depleted cells and by 8 h there was no AP-1 DNA-binding activity. However, when IL-2 was present there was no loss of AP-1 DNA-binding activity and no apoptosis. Therefore, in the presence of IL-2, AP-1 levels remain at a sufficiently high level to ensure cell cycle progression. Phorbol esters which, as described above, induce gene transcription through the TPA-responsive element can induce AP-1 DNA-binding activity in IL-2-depleted cells which have lost their AP-1 DNA-binding activity. However, unlike the form of AP-1 induced by IL-2, the AP-1 induced by TPA was still sensitive to gluco-corticoids. In the presence of the steroid AP-1, DNA-binding activity was rapidly lost and the cells underwent apoptosis.

A tyrosine kinase is likely to catalyse the additional modifications to the components of the AP-1 transcription factor complex to confer resistance to glucocorticoids. At present this tyrosine kinase-mediated signal has not been characterised. It has been shown by Pulverer et al. (1991) that the transactivating ability of c-*Jun* is stimulated by MAP kinase. The MAP kinase family of serine/threonine kinase is activated by phosphorylation on either serine/threonine residues and/or tyrosine residues. Since the 42-kDa MAP-2 kinase is tyrosine phosphorylated by the lymphocyte-specific tyrosine kinase p56^{lck}, as described above, and by PKC (Nel et al. 1991) the IL-2-stimulated tyrosine phosphorylation step may well affect the DNA-binding activity of AP-1 and alter its ability to interact with the GR by protein–protein interactions.

6.7 Susceptibility to Apoptosis and Sensitivity to Chemotherapeutic Agents

Cells that are resistant to undergoing apoptosis in the presence of one stimulus are not necessarily resistant to all stimuli because of the exist-ence of multiple mechanisms for triggering apoptosis. However, as described above, in those cells in which *bcl-2* is overexpressed, they can acquire simultaneous resistance to growth factor withdrawal and che-motherapeutic agents. To assess whether the MT-transformed cells were more resistant to apoptosis in the presence of chemotherapeutic drugs we challenged these cells with the DNA-damaging, topoisomerase II inhibitor VM26. This drug has been shown to be a powerful inducer of apoptosis in a number of cells (Walker et al. 1991). Indeed, cells, in general, have been found to be very sensitive to DNA damage. All of the T51B cells in this study underwent apoptosis in the presence of VM26 with the exception of the MT-transformed cells (Fig. 3b). The p53 tumour suppressor gene appears to play a role in the detection of DNA damage (Kastan et al. 1991). Apoptosis is triggered in those cells that cannot repair their DNA by a p53-mediated mechanism. All the T51B-derived cells expressed the p53 oncogene. Thus, the resistance of these cells to apoptosis could not be overriden by the DNA damage-p53 detection system that is capable of inducing apoptosis in the other cell lines. These results show that mutations that decrease the sensitivity of cells to apoptosis following serum or growth factor withdrawal can, in some circumstances, lead to an across-the-board resistance to other apoptosis-inducing agents. The *bcl-2* oncogene is a prime example of this. Moreover, in the case of MT, the resistance is conferred by a *bcl-2*-independent mechanism.

6.8 Conclusion

These studies clearly demonstrate that proto-oncogenes play a role in the control of apoptosis (Fig. 4). Oncogenes that are now implicated include *fos, jun, myc*, p53, *bcl-2, ras, src* and possibly *fyn, yes* and p56^{lck}. Just as the identification of oncogenes contributed to our understanding of regulation of the cell cycle the implication of these oncogenes in apoptosis reveals details of the control of apoptosis and its relationship to the cell cycle.

There are two principal triggers of apoptosis in vivo. The first is a tissue- or cell-type-specific signal to remove individual cells from the tissue or cell population and the second a decision, seemingly made intracellularly, that a damaged cell should be eliminated to avoid its contribution to an altered phenotype. In the first case it typically results

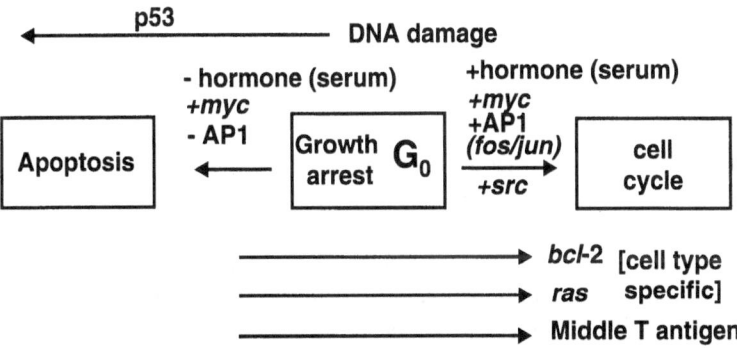

Fig. 4. Summary of the role of oncogenes in the control of apoptosis

from withdrawal of trophic factor and causes the cell to move from whatever state it is in towards apoptosis. Thus, cycling cells exit the cell cycle and differentiated cells cease expression of differentiated function and go directly into apoptosis. We do not know to what extent movement along the pathway of apoptosis is reversible. From our studies on CTLL cells they can be rescued from cell death if IL-2 is added back within 4 h, but if they are left 6–8 h then most of the cells cannot be rescued. Significantly, even after prolonged incubation without IL-2 some cells can be rescued following IL-2 readdition, indicating that a small fraction of any population may be quite resistant to growth factor withdrawal. There are also hormones or cytokines such as transforming growth factor β or tumour necrosis factor which directly target certain cells and induce cell death. Two families of intracellular signals mediated repression of apoptosis and promote re-entry into the cell cycle. PKC appears to repress cell death in cooperation with tyrosine kinases. The *src* family of kinases appears to be involved and the role played by each kinase appears cell type-specific. These kinases directly couple apoptosis to the cell cycle. Viral genes, such as MT, that can exploit this relationship by permanently activating these kinases can very efficiently transform cells by simultaneously repressing apoptosis and driving cell cycle progression. Other protooncogenes such as *ras* and *myc* are also involved in driving cell cycle repression, although the role of *myc* which can also activate apoptosis and even affect the course of differentiation in some cells is far from clear. The expression of sufficient *fos* and *jun* to form AP-1 is essential for cells to remain in the cell cycle and avoid going into apoptosis.

Cells are extremely sensitive to damage to their DNA and p53 is involved in detection of this damage and, failing its repair, signalling apoptosis. Alterations in p53 can lead to a failure to signal adequate DNA repair and a failure to undergo apoptosis, thereby contributing to the malignant phenotype. This is an example of how a gene that is not directly involved in the cell cycle or apoptotic pathway, but is involved in the detection of cellular damage can contribute to tumorigenesis. As we learn more about the different sensors of cellular damage and the transducers of this information into a cellular response we are likely to uncover other potential oncogenes.

Finally, although the function of the *bcl*-2 gene product has not been established, it appears unique in the sense that it does not appear to participate in either cell cycle progression or damage detection (although an involvement in oxidative stress cannot be ruled out); this gene may represent an apoptosis-specific gene. The existence of such a gene as well as the operation of other mechanisms for repressing apoptosis can interfere with drugs designed to exploit the cell death pathway. As we learn more about these mechanisms combination therapies involving growth-factor depletion and apoptosis-inducing drug are likely to emerge.

References

Allsop, T. E., Wyatt, S. Paterson, H. F. Davies, A. M. (1993) The proto-on-cogene bcl-2 can selectively rescue neurotrophic factor-dependent neurons from apoptosis. Cell 73:295–307

Amati B, Littlewood, TD, Evan GI, Land H (1993) The c-myc protein induces cell cycle progression and apoptosis through dimerization with max. EMBO J 12:5083–5087

Angel P, Baumann I, Stein B, Delius H, Rahmsdorf HJ, Herrlich P (1987) 12-o-tetradecanoyl-phorbol-13-acetate induction of the human collagenase gene is mediated by an inducible enhancer element located in the 5'-flanking region. Mol Cell Biol 7:2256–2266

Araki S, Shimada Y, Kaji K, Hayashi H (1990a) Apoptosis of vascular endo-thelial cells by fibroblast growth factor deprivation. Biochem Biophys Res Commun 168:1194–1200

Araki S, Shimada Y, Kaji K, Hayashi H (1990b) Role of protein kinase C in the inhibition by fibroblast growth factor of apoptosis in serum-depleted endothelial cells. Biochem Biophys Res Commun 172:1081–1085

Arends MJ, McGregor AH, Toft NJ, Brown EJH, Wyllie AH (1993) Suscepti-bility to apoptosis is differentially regulated by c-myc and mutated Ha-ras oncogenes and is associated with endonuclease availability. Br J Cancer 68:1127–1133

Ascaso R, Marvel J, Collins MKL, Lopez-Rivas A (1994) Interleukin-3 and bcl-2 cooperatively inhibit etoposide-induced apoptosis in a murine pre-B cell line. Eur J Immunol 24:537–541

Bennett MR, Evan GI, Newby AC (1994) Deregulated expression of the c-myc oncogene abolishes inhibition of proliferation of rat vascular smooth muscle cells by serum reduction, interferon, heparin, and cyclic nucleotide analogues and induces apoptosis. Circ Res 74:525–536

Diamond MI, Miner JN, Yoshinaga SK, Yamamoto KR (1990) Transcription factor interactions: positive or negative regulation from a single DNA ele-ment. Science 249:1266–1272

Fisher TC, Milner AE, Gregory CD, Jackman AL, Aherne GW, Hartley JA, Dive C, Hickman JA (1993) bcl-2 modulation of apoptosis induced by anticancer drugs: resistance to thymidylate stress is independent of classical resistance pathways. Cancer Res 53:3321–3326

Gillis S, Smith KA (1977) Long term culture of tumor-specific cytotoxic T cells. Nature 208:154–156

Gourdeau H, Walker PR (1994) Evidence for trans regulation of apoptosis in intertypic somatic cell hybrids. Mol Cell Biol 14:6125–6134

Guy CT, Muthuswamy SK, Cardiff RD, Soriano P, Muller WJ (1994) Activation of the c-Src tyrosine kinase is required for the induction of mammary tumors in transgenic mice. Genes Dev 8:23–32

Hennet T, Bertoni G, Richter C, Peterhaus E (1993) Expression of BCL-2 protein enhances the survival of mouse fibrosarcoid cells in tumor necrosis factor-mediated cytotoxicity. Cancer Res 53:1456–1460

Hockenbury DM, Zutter M, Hickey W, Nahm M, Korsmeyer S (1991) Bcl2 protein is topographically restricted in tissues characterized by apoptotic cell death. Proc Natl Acad Sci USA 88:6961–6965

Kanter P, Leister KJ, Tomei LD, Wenner PA, Wenner CE (1984) Epidermal growth factor and tumor promoters prevent DNA fragmentation by different mechanisms. Biochem Biophys Res Commun 118:392–399

Kastan MB, Onyekwere O, Sidransky D, Vogelstein B, Craig RW (1991) Participation of p53 in the cellular response to DNA damage. Cancer Res 51:6304–6311

Lowe SW, Jacks T, Housman DE, Ruley HE (1994) Abrogation of oncogene-associated apoptosis allows transformation of p53-deficient cells. Proc Natl Acad Sci USA 91:2026–2030

McConkey DJ, Hartzell P, Jondal M, Orrehius S (1989) Inhibition of DNA fragmentation in thymocytes and isolated thymocyte nuclei by agents that stimulate protein kinase C. J Biol Chem 264:13399–13402

McDonnell TJ, Troncoso P, Brisbay SM, Logothetis C, Chung LWK, Hsieh J-T, Tu S-M, Campbell ML (1992) Expression of the protooncogene bcl-2 in the prostate and its association with emergence of androgen-independent prostate cancer. Cancer Res 52:6940–6944

McDonnell TJ, Marin MC, Hsu B, Brisbay SM, McConnell K, Tu S-M, Campbell ML, Rodriguez-Villanueva J (1993) The bcl-2 oncogene: apoptosis and neoplasia. Radiat Res 136:307–312

Miner JN, Diamond MI, Yamamoto KR (1991) Joints in the regulatory lattice: composite regulation by steroid receptor-AP1 complexes. Cell Growth Differ 2:525–530

Munoz E, Zubiaga AM, Huber BT (1991) Tyrosine protein phosphorylation is required for protein kinase C-mediated proliferation in T cells. FEBS Lett 279:319–322

Nel AE, Hanekom C, Hultin L (1991) Protein kinase C plays a role in the induction of tyrosine phosphorylation of lymphoid microtubule-associated protein-2 kinase. J Immunol 147:1933–1939

Oitavi ZN, Milliman CL, Korsmeyer SJ (1993) Bcl-2 heterodimerizes in vivo with a conserved homolog, Bax, that accelerates programmed cell death. Cell 74:609–619

Oliver FJ, Marvel J, Collins MKL, Lopez-Rivas A (1993) Bcl-2 oncogene protects a bone marrow-derived pre B cell line from 5'-fluor, 2'-deoxyuridine-induced apoptosis. Biochem Biophys Res Commun 194:126–132

Ponta H, Cato ACB, Herrlich P (1992) Interference of pathway specific transcription factors. Biochim Biophys Acta 1129:225–261

Pulverer BJ, Kyriakis JM, Avruch J, Nikolakaki E, Woodgett JR (1991) Phosphorylation of c-jun mediated by MAP kinases. Nature 353:670–674

Riabowol K, Schiff J, Gilman MZ (1992) Transcription factor AP-1 activity is required for initiation of DNA synthesis and is lost during cellular aging. Proc Natl Acad Sci USA 89:157–161

Rodriguez-Tarduchy G, Lopez-Rivas A (1989) Phorbol esters inhibit apoptosis in IL-2 dependent T lymphocytes. Biochem Biophys Res Commun 164:1069–1075

Roy C, Brown DL, Little JE, Valentine BK, Walker PR, Sikorska M, Leblanc J, Chaly N (1992) The topoisomerase II inhibitor teniposide (VM26) induces apoptosis in unstimulated mature lymphocytes. Exp Cell Res 200:416–424

Royal I, Gourdeau H, Marceau N (1992) Down-regulation of cytokeratin 14 mRNA in polyoma virus middle T-transformed rat liver epithelial cells. Cell Growth Differ 3:589–596

Ryseck R-P, Bravo R (1991) c-Jun, junB and junD differ in their binding affinities to AP-1 and CRE consensus sequences: effect of Fos proteins. Oncogene 6:533–542

Smith KA (1989) The interleukin 2 receptor. Annu Rev Cell Biol 5:397–425

Thomas JE, Aguzzi A, Soriano P, Wagner EF, Brugge JS (1993) Induction of tumor formation and cell transformation by polyoma middle T antigen in the absence of src. Oncogene 8:2521–2529

Tomei LD, Kanter P, Wenner CE (1988) Inhibition of radiation-induced apoptosis in vitro by tumor promoters. Biochem Biophys Res Commun 155:324–331

Walker PR, Smith C, Youdale T, Leblanc J, Whitfield JF, Sikorska M (1991) Topoisomerase II-reactive chemotherapeutic drugs induce apoptosis in thymocytes. Cancer Res 51:1078–1085

Walker PR, Kwast-Welfeld J, Gourdeau H, Leblanc J, Neugebauer W, Sikorska M (1993a) Relationship between apoptosis and the cell cycle in lymphocytes: roles of protein kinase C, tyrosine phosphorylation and AP1. Exp Cell Res 207:142–151

Walker PR, Kwast-Welfeld J, Sikorska, M. (1993b) Relationship between apoptosis and the cell cycle. In: Lavin M, Watters D (eds) Programmed cell death. Harwood Academic, Chur, Switzerland

Whitfield JF (1990) Calcium, cell cycles and cancer. CRC Press, Boca Raton, Florida

Williams GT (1991) Programmed cell death: apoptosis and oncogenesis. Cell 65:1097–1098

7 Tenascin and Extracellular Matrix: Possible Biological Implications During Regression and Carcinogenesis of the Prostate

G. Vollmer, S. Schenk, and H. Michna

7. 1 Introduction

The role of extracellular matrices (ECM) in normal prostate function during the regression of the prostate and in carcinogenesis of this organ has only recently attracted the attention of various research groups. Histologically, the most prominent structure of the ECM in the prostate is the basement membrane (BM). This structure deserves particular attention, since it exhibits unique features in the prostate, not observable in other organs. In the fetal and adult prostate and in various hyperplastic conditions the BM was found to be continuous with local thickenings and therefore rather inconspicious compared to the BM of other organs (Bonkhoff et al. 1991). However, whereas adenocarcinoma cells of most other organs degrade and invade the BM after having acquired their invasive phenotype, prostatic tumor cells, independent of their degree of differentiation, persist with their pronounced BM (Bonkhoff et al. 1991, 1992). Even highly malignant anaplastic and small cell carcinomas, irradiated and/or hormonally treated tumors, as well as lymphatic or hematogenous mestastases showed distinct BM formations in contact with the stroma (Bonkhoff et al. 1992). These observations suggest a close association and a strong dependency of the normal and carcinogenic prostatic epithelium for BM anchoring. Experimental evidence in favor of this assumption has been delineated from in vitro findings in cultured primary human prostatic epithelial cells. For these cells it has been demonstrated that a reconstituted BM (Matrigel) promotes morphological and functional differentiation in vitro (Fong et al. 1991). In vitro experiments further suggested that this reconstituted BM is able to stimulate the growth of human prostatic carcinoma in athymic mice and to control the migration of these tumor cells in cell culture (Passaniti et al. 1992). In this context it is important to mention that highly metastatic prostate tumor cells could be selected out of a wild-type tumor cell line using a Matrigel invasion assay. Analysis of the cell surface integrins showed that the metastatic phenotype exhibited down-regulation of the $\alpha_3\beta_1$ integrin and an overexpression of the $\alpha_6\beta_4$ integrin. Further, whereas the α_6-subunit was complexed to β_1 subunit in wild-type cells (Dedhar et al. 1993), it was predominantly associated with β_4 in the metastatic cells. Alterations in the expression of the $\alpha_3\beta_1$ and $\alpha_6\beta_4$ integrins may thus allow these cells to become more invasive and lead to an increased propensity for metastasis.

In contrast to the striking BM dependency of normal and malignant prostatic epithelial cells, apoptotic prostate epithelial cells apparently not only lose their BM anchoring, but after entering the process of activated cell death, the ECM underlying the dying cells is degraded. Whereas matrix degradation obviously requires the activation of proteases (for review see Tenniswood et al. 1992), the molecular events triggering the detachment of the cells remain obscure. Recently, two papers have implicated integrin-mediated signaling as a controlling factor for apoptosis of epithelial cells. Meredith et al. (1993) proposed that the ECM serves as a survival factor for human endothelial cells, which immediately enter a suicidal pathway if they lack cell–matrix interactions. Frisch and Francis (1994) have named this phenomenon anoikis, a term derived from the Greek word "homeless." The anoikis phenomenon implies that once a differentiated cell loses contact with its underlying matrix, it dies. This hypothesis is in line with findings derived from another experimental system. With an anti-β_2 integrin (complement receptor type 3) antibody, apoptosis could be induced in microglia (Reid et al. 1993). Since tumor cells can gain a selective growth advantage by blocking the apoptotic process (Reed 1994), the question has recently been raised (Ruoslahti and Reed 1994) as to whether the attachment of cells to the matrix mediated by any integrin will be sufficient to abrogate anoikis or whether several integrins will have to bind to enable a given cell to survive and to proliferate. The outcome of this experimental work provides evidence for the role of some integrins in preventing normal epithelial cells from entering the apoptotic process.

Another function of integrins in the apoptotic process has long been identified: integrins of macrophages, in particular, play a key role in removing apoptotic cells, namely, the vitronectin receptor ($\alpha_v\beta_3$-integrin; Savill et al. 1990, Fadok et al. 1992).

Normal development and function as well as carcinogenesis of hormone-dependent glandular organs, particularly of the prostate, also appear to be dependent on functional interactions between the tissue compartments epithelium and stroma. In fact, the mesenchyme serves as the major androgen target and determines the fate of the epithelium (Cunha and Donjacour 1987; Cunha et al. 1987). This raises the question of the molecular nature of the signals which are provided by the mesenchymal stroma. One signal transduction pathway may be through the release of soluble paracrine-acting factors. Although numerous growth factors including prostatrophin, basic fibroblast growth factor (bFGF), and epidermal growth factor (EGF) are expressed and probably functional in the prostate (McKeehan 1991; Story 1991; Wilding 1991; Crabb et al. 1986 a,b), no real candidate molecule has been identified so far that might serve as the paracrine signal between the stroma and the epithelium in the prostate.

In addition to soluble factors, ECM molecules or various combinations of them may participate in reciprocal stromal–epithelial interactions. A candidate molecule is the glycoprotein tenascin. Tenascin/hexabrachion protein is a significant ECM glycoprotein which has been isolated and variably characterized by several research groups (for review see Erickson and Lightner 1988, Erickson and Bourdon 1989). Initially, Chiquet-Ehrismann et al. (1986) proposed that tenascin expression is intimately correlated to epithelial–mesenchymal interactions during fetal development and oncogenesis. With respect to quantitative aspects of tenascin expresssion, this conclusion still holds. In contrast to earlier observations, low levels of tenascin are expressed in normal adult organs, including skin (Schalwijk et al. 1991), human placenta (Castellucci et al. 1991), human liver (Ramadori et al. 1991), and human and rodent breasts (Inaguma et al. 1988; Ferguson et al. 1990; Howeedy et al. 1990; Koukoulis et al. 1991). In experimental skin wounds, a conspicuous increase in tenascin expression was shown in the BM region of the regenerating epidermis as well as in the subjacent granulation tissue (Mackie et al. 1988). In view of all these observations, tenascin appears to be a valuable marker glycoprotein to visualize stromal reactions on epithelial inputs both in normal and malignant glandular epithelial tissue, as has previously been shown for breast, endometrium, and prostate (Mackie et al. 1987; Inaguma et al. 1988; Howeedy et al. 1990; Koukoulis et al. 1991; Vollmer et al. 1990, 1991, 1992; Ibrahim et al. 1993).

In this paper we examine tenascin expression in normal prostate glands and in prostates containing malignant tissue in humans and rodents. We further describe the effect of androgen ablation on tenascin expression in regressing normal prostates and regressing Dunning R 3327 H prostate carcinoma and comment on the influence of tumorigenesis on serum tenascin levels. Finally, we discuss possible functions of tenascin in various normal and malignant conditions of the prostate.

7.2 Tenascin in Human Prostate and Prostate Carcinoma

Tissue sections of normal and malignant human prostate were stained by a polyclonal anti-tenascin antibody (provided by Prof. H.P. Erickson, Duke, University, Durham, USA; for reference see Lightner et al. 1989). In the normal differentiated human prostate we found little, if

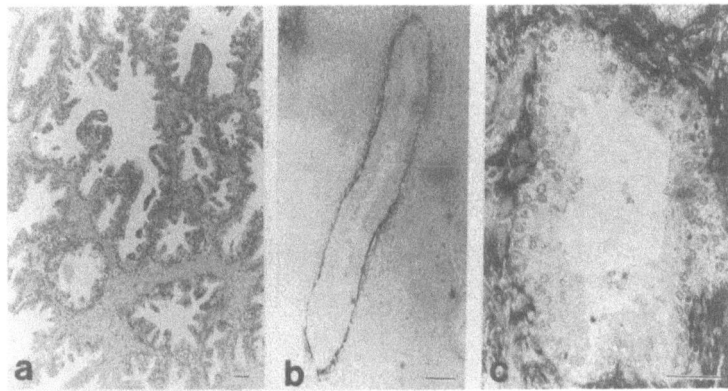

Fig. 1a–c. Tenascin in normal, hyperplastic and malignant human prostate. Paraffin sections of normal (**a**), hyperplastic (**b**), and malignant (**c**) human prostates were analyzed immunohistochemically using a polyclonal anti-tenascin antiserum. *Calibration bars* represent 100 μm

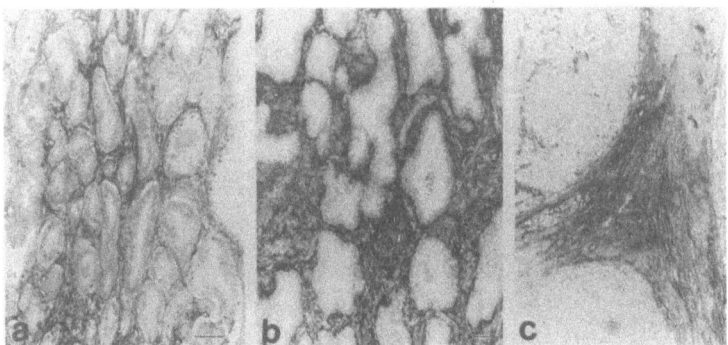

Fig. 2a–c. Tenascin in adenocarcinomas of the human prostate. Paraffin sections of highly differentiated (**a**), moderately differentiated (**b**), and poorly differentiated (**c**) adenocarcinoma of the human prostate were stained with a polyclonal anti-tenascin antibody. *Calibration bars* represent 100 μm

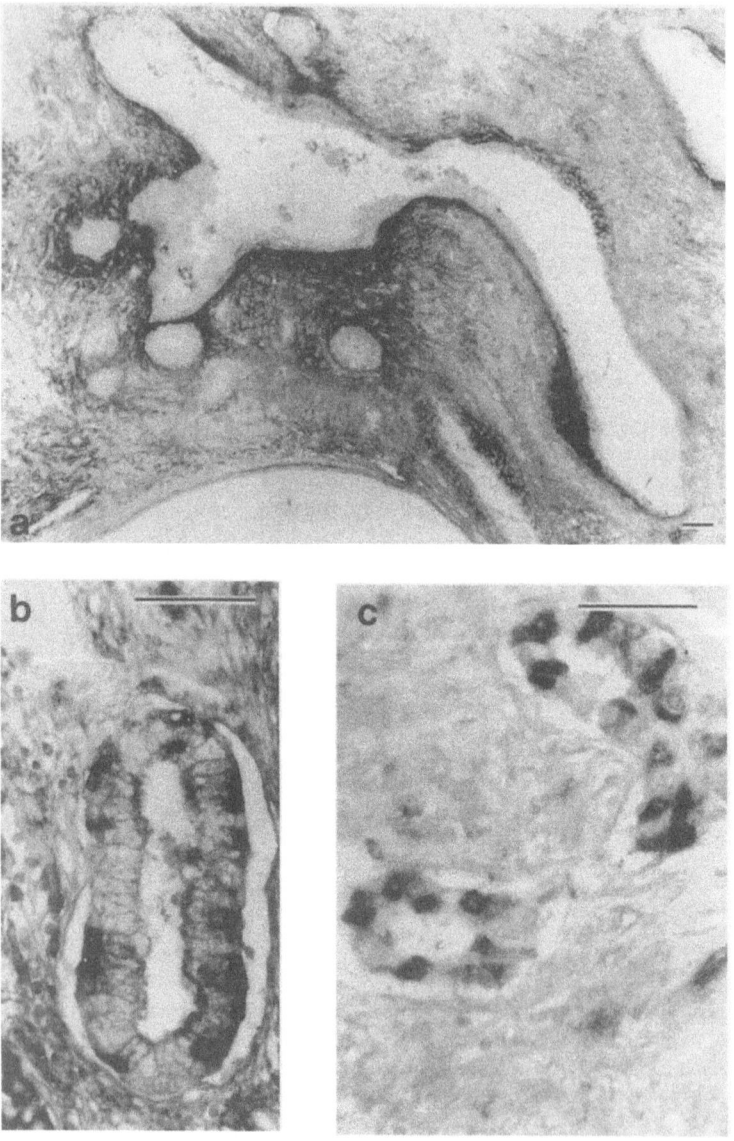

Fig. 3a–c. Legend see p. 129

any, tenascin immunoreactivity (Fig. 1a). Following the course of carcinogenesis, tenascin expression was enhanced in hyperplastic tissues. In hyperplastic sections we found reaction product predominantly associated with BM structures (Fig. 1b). As in all other tumors investigated so far, tenascin immunoreactivity covered almost the entire ECM of the stromal mesenchyme in prostatic adenocarcinoma (Fig. 1c, Fig. 2), and as in cases of human endometrial adenocarcinoma (Vollmer et al. 1990), the localization and the staining intensity appeared to be independent of the differentiation of the tumor (Fig. 2). However, in two specimens of human prostatic carcinoma we observed an immunoreaction within the epithelial tissue compartment. The finding of epithelial-derived tenascin is of particular interest, since, initially, tenascin expression has been described to be restricted to mesenchymal tissue compartments (Fig. 3). With the more sensitive methods such as in situ hybridization, northern blotting, or reverse transcription polymerase chain reaction (rtPCR) approaches more and more evidence is accumulating whichs indicate that developing epithelia (Koch et al. 1991), normal epithelia, and, particularly, malignant adenocarcinoma cells, including those of the prostate (Ibrahim et al. 1993), the breast (Lightner et al. 1994) and the endometrium (Vollmer, unpublished observations), are able to produce tenascin. However, the biological significance of epithelial-derived tenascin remains to be elucidated.

◀ **Fig. 3a–c.** Epithelial-derived tenascin in human prostate tissue. In paraffin sections of two human prostate cancer specimen, epithelial-derived tenascin could be identified. One specimen is shown in an overview representation (**a**), the second at a higher magnification (**b,c**). *Calibration bars* represent 100 μm

7.3 Tenascin Expression in the Prostate
and Prostatic Tumor Models of the Rat

7.3.1 Tenascin Expression in the Normal
and Regressing Rat Prostate

Tenascin expression was examined immunohistochemically in the normal rat prostate and in the regressed prostate following androgen ablation. In two series of experiments intact and orchiectomized animals were subjected to treatment with the antiandrogens flutamide (1 mg/animal per day), casodex (1, 3, or 10 mg/animal per day), and cyproterone acetate (CPA; 3 mg/animal per day). Untreated animals and orchiectomized animals were used as controls. After a treatment period of 14 days we observed little, if any, tenascin immunoreactivity in untreated control animals (Fig. 4a). Androgen deprivation both by orchiectomy (not shown) or antiandrogen treatment (Fig. 4b-d) resulted in tenascin expression in the stroma of the regressing prostates. The observed staining patterns were variable both in intensity and in area covered, except for CPA treatment, for which relatively uniform staining patterns have been observed (Fig. 4d).

These results demonstrate that androgen ablation induces expression of tenascin and a large body of evidence already exists indicating that the directed interaction of the epithelium with the stroma of various developing organs or during the oncogenic process play a crucial role in tenascin expression (Chiquet-Ehrismann et al. 1986; Inaguma et al. 1988; Aufderheide et al. 1987; Aufderheide and Ekblom 1988). Our data provide evidence to support the idea that both epithelial–mesenchymal interactions and remodeling of the extracellular matrix represent processes that are associated with the involution of the prostate. The latter apparently occurs relatively late in the apoptotic cascade, probably after the apoptotic process has been completed.

Fig. 4a–d. Induction of tenascin expression in the regressed rat prostate. Tenascin expression was examined by immunohistochemistry in the normal rat prostate (**a**) and in the regressed prostate following androgen ablation. We show tenascin expression of intact animals that were subjected to treatment with flutamide (**b**), casodex (**c**), or cyproterone acetate (**d**). *Arrows,* tenascin reaction product

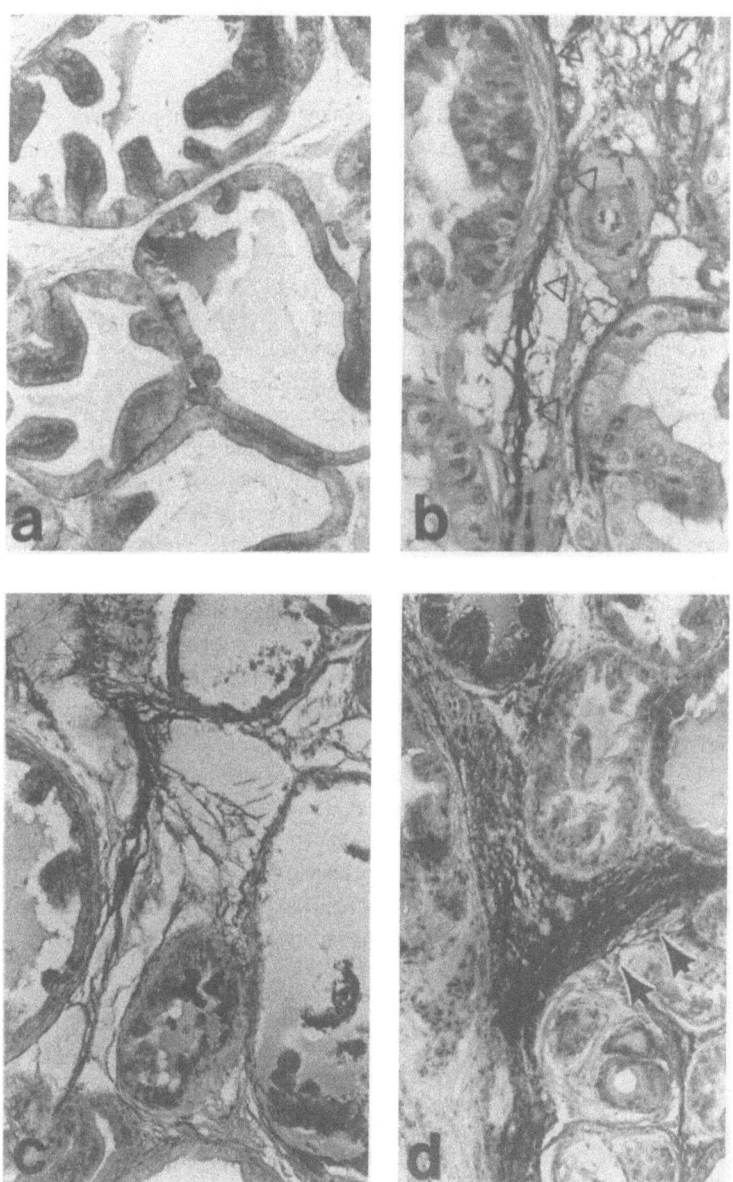

Fig. 4a–d. Legend see p. 130

7.3.2 Tenascin Expression
in the Dunning R 3327 H Prostate Carcinoma

Experimentally, Dunning R 3327 H prostate adenocarcinoma tissue was subcutaneously inoculated into the hind limbs of male rats. Thirteen weeks after inoculation rats were either orchiectomized or received a daily dose of 3 mg/animal per day of either CPA or flutamide. The various treatments differentially affected tumor growth. Orchiectomy reduced tumor growth by 60% compared to untreated controls. Antiandrogens were less effective but reduced tumor weight by approximately 30% compared to the untreated control tumors.

Despite the significant effects of androgen ablation on growth of the Dunning tumor, the staining pattern of tenascin immunoreactivity was unaffected by the hormonal treatment. In both treatment groups tenascin-like immunoreactivity stained the extracellular space of the stromal mesenchyme, the staining intensity depending on the morphology of the tumor. In those tumors with small ducts and little stroma, tenascin immunoreactivity was moderate (Fig. 5a). In those tumors with enlarged ducts and a significant stromal tissue compartment, the tumors were highlighted by an intense stromal reaction for tenascin (Fig. 5b). In conclusion, the Dunning tumor can be added to the list of adenocarcinomas which exhibit an upregulation of tenascin expression as the result of tumorigenesis. Further, these results clearly demonstrate that tumorigenesis disrupts the hormonal regulation of tenascin expression.

In two cases of untreated Dunning tumors we observed tenascin immunoreactivity in the epithelial tissue compartment (Fig. 5c). The considerations for the possible significance of epithelial-derived tenascin are basically the same as for epithelial-derived tenascin in human prostate carcinomas and have been intensively discussed above.

Fig. 5a–c. Tenascin in the Dunning R 3327 H rat prostate carcinoma. Tissue ▶ sections were stained with a polyclonal anti-tenascin antibody. Morphologically, these tumors were either composed of small ducts with little stroma (**a**) or by enlarged glands with a significant stromal tissue compartment (**b**). In two specimens epithelial-derived tenascin could be identified (**c**). *Calibration bars* represent 100 μm. *Arrows,* tenascin reaction product

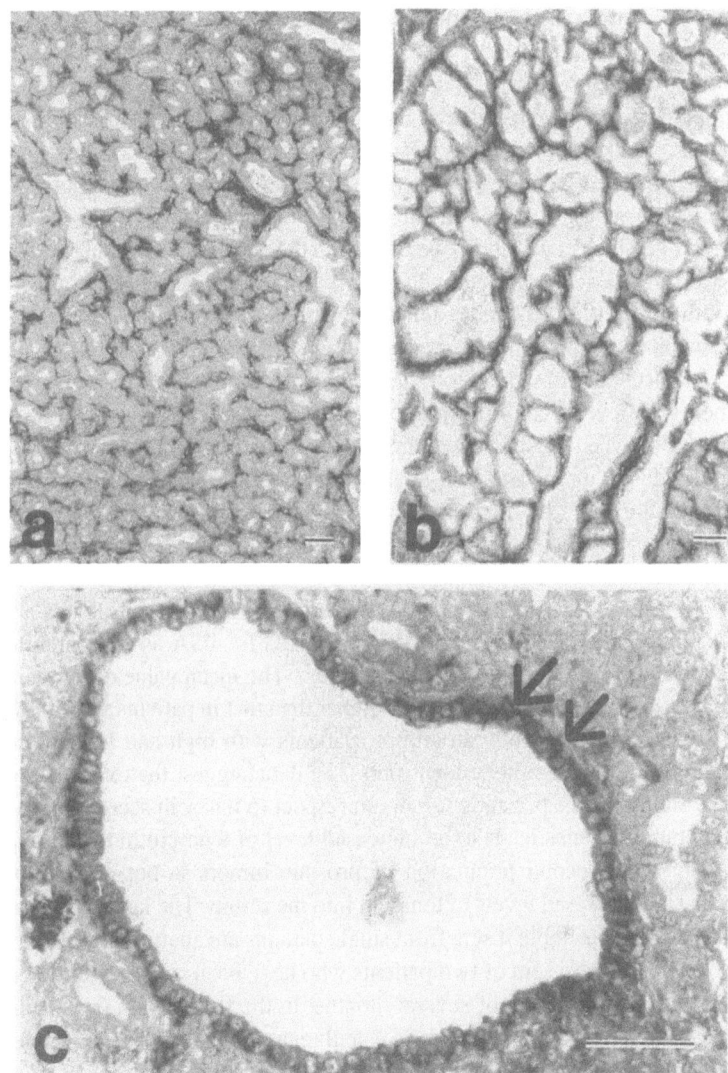

Fig. 5a–c. Legend see p. 132

7.4 Is Tenascin a Tumor Marker for Prostate Carcinoma?

The strong expression of tenascin immunoreactivity in prostate tumors of humans and rodents, as well as in all other tumor species investigated so far, suggested to us that tenascin might be a tumor marker, provided that it is released into the serum. Since there were recent reports of increased tenascin levels in the sera of tumor patients (Herlin et al. 1991; Washizu et al. 1993), we addressed the question of whether serum levels of tenascin correlate with the occurrence of prostate carcinomas. With an ELISA procedure tenascin was measured in the sera of 47 patients in whom prostate-specific antigen (PSA) levels had routinely been checked. For evaluation these sera were subgrouped according to their PSA levels. The first subgroup contained those samples with PSA values between 0.2 and 10 ng/ml (17 normal PSA values between 0.2 ng/ml and 4 ng/ml and two slightly elevated samples between 4 and 10 ng/ml PSA). The second subgroup comprised all those samples with clinically suspicious PSA levels above 10 ng/ml. As a control group we used the sera of ten pregnant women.

As can be seen in Fig. 6 mean values of serum tenascin levels are increased in patients who have been checked for PSA as compared to tenascin levels in healthy pregnant women. The mean value of tenascin in patients with high PSA levels is higher than that in patients with a low PSA. However, in both subgroups, patients with high and low serum tenascin levels have been identified. The data suggest the existence of two subgroups of prostatic tumors in respect to tenascin secretion. The first tumor species leads to an increased level of tenascin in the serum, whereas the second population of prostate tumors is not capable of releasing increased levels of tenascin into the serum. The latter observation is also detectable if sera from single patients are analyzed over time. We had access to sera of two patients who have been routinely checked for PSA over a period of several months. In the first patient (shown in Fig. 6b) a correlation between PSA and tenascin levels was detectable, which was not observed in the serum of patient 2. As can be seen from Fig. 6c, tenascin levels in the serum of this patient remain almost constant although serum PSA levels increase. There is no mechanistic explanation at all for the fact that some tumors seem to release tenascin into the serum while others do not; this deserves further investigation.

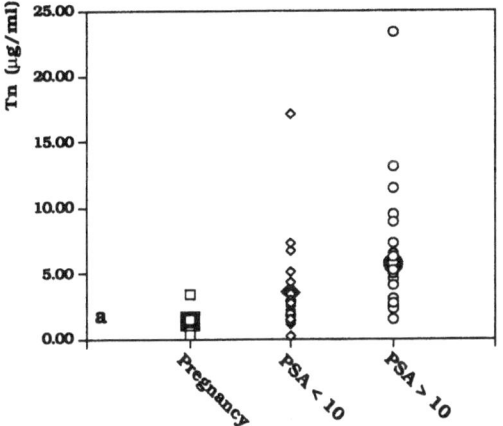

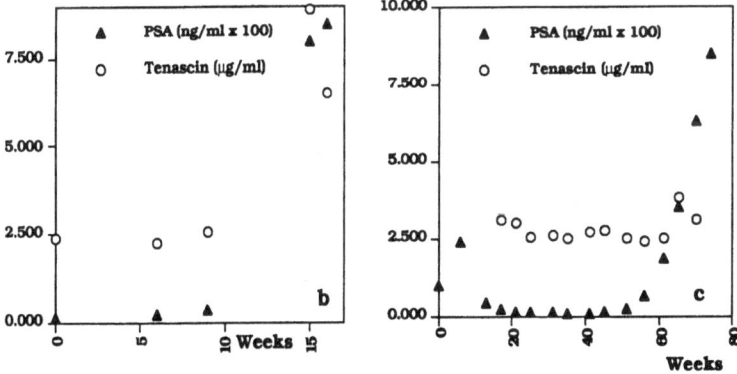

Fig. 6a–c. Serum tenascin levels in prostate cancer patients. Forty-seven patients that had been routinely checked for their *PSA* levels were analyzed for serum tenascin levels. We show tenascin levels (*open symbols*) for patients with low PSA levels and for patients with high PSA levels. As control, sera from healthy pregnant women were used. *Solid symbols* indicate mean values of each group (**a**). Additionally, two patients were checked for PSA over time. In the serum of one patient tenascin levels paralleled increasing PSA levels (**b**); in the serum of the other patient no correlation was detectable (**c**)

These data on tenascin levels in sera of prostate cancer patients clearly indicate that there is no convincing correlation between serum tenascin levels and the existence of prostate carcinomas. These data further suggest that tumorigenesis can be regarded as a potential reason for increased serum tenascin levels; however, these data also demonstrate that many tumors do not release tenascin into the serum. The elevated tenascin values of some patients with a normal PSA level of less than 10 ng/ml clearly argue for potential pathological conditions in addition to carcinogenesis in which serum tenascin levels are elevated. Only recently has evidence been provided that tenascin might be an acute phase protein and that infectious and inflammatory conditions might be the main reason for elevated tenascin levels in the serum (Schenk, unpublished observations).

7.5 Involvement of Tenascin in Reciprocal Epithelial–Mesenchymal Interactions

The almost exclusive localization of tenascin in the mesenchymal (stromal) compartments of developing organs and organs undergoing oncogenesis prompted many investigators to hypothesize that tenascin reflects a stromal reaction to epithelial morphogenesis or oncogenesis. In embryonic organs it has been clearly demonstrated that the epithelium induces tenascin expression in the mesenchymal tissue compartment (Aufderheide et al. 1987; Aufderheide and Ekblom 1988; Ekblom and Aufderheide 1989). These studies further implied that tenascin expression may be induced by soluble epithelial-derived growth factors. This hypothesis has been proven in many in vitro experiments. Using fibroblasts or other mesenchymal-derived cells as target, various growth factors have been identified as inducers for tenascin (for summary see Table 1). Our own results with human endometrial stromal cells suggest that conditioned media from endometrial tumor cells and the growth factors transforming growth factor-β and insulin-like growth factor-I have the potency to induce enhaced tenascin mRNA expression.

Table 1. Growth factors stimulate tenascin expression

Growth factor cytokine	Tissue/cell line	Reference
Serum	Chicken skin fibroblasts	Pearson et al. 1988; Chiquet Ehrismann et al. 1989
TGF-β	Fetal skin fibroblasts	Pearson et al. 1988; Chiquet Ehrismann et al. 1989
	Vascular smooth muscle cells	Mackie et al. 1992
	Liver smooth muscle cells	Schwogler et al. 1992
	Swiss 3T3	Tucker et al. 1993
	Endometrial stroma cells	Vollmer, unpublished observations
NGF/monosialogangliosides	CG glioma cells	Yavin et al. 1991
Interleukin-1	Synovial fibroblasts	McCachren et al. 1992
PDGF	Vascular smooth muscle cells	Mackie et al. 1992
Angiotensinogen II	Vascular smooth muscle cells	Mackie et al. 1992
bFGF	Swiss 3T3	Tucker et al. 1993
IGF-I	Endometrial stroma cells	Vollmer, unpublished observations

TGF-β, transforming growth factor; NGF, nerve growth factor; PDGF, platelet-derived growth factor; bFGF, basic fibroblast growth factor; IGF-I, insulin-like growth factor-I.

7.6 Summary and Conclusions

In this paper we have described the expression of the glycoprotein tenascin in normal and malignant prostates of humans and rats. We also reported on its enhanced expression in the regressed prostate following androgen ablation. In our view tenascin does not represent a marker for prostate tumors and carcinogenesis of the prostate in general; instead, it has to be regarded as a valuable marker glycoprotein that can be used to understand various aspects of the biochemical mechanisms underlying regression and carcinogenesis of the prostate. Since it is mainly induced by directed interactions of the epithelium with the stroma, it might be particularly valuable in assessing the potential importance of epeithelial–mesenchymal interactions in the processes of regression and tumorigenesis of the prostate. To discover the ultimate role of tenascin in the regressing prostate it is important to investigate whether tenascin represents a glycoprotein that actively participates in the apoptotic process or whether its expression is a result of tissue remodeling once the elimination of cells by apoptosis is completed. Our studies on tenascin expression during regression and oncogenesis of the prostate clearly demonstrate the involvement of the ECM in both processes. The detailed knowledge of molecular mechanisms regulating the expression, remodeling, structure, and composition of the ECM and their interaction with cells may provide new clues for the design and delivery of hormonal and nonhormonal therapies for prostate cancer.

Acknowledgements. The authors would like to thank Ms. Kirsten Ebert for her excellent technical assistance. This paper was funded by the Deutsche Forschungsgemeinschaft SFB 367/A6.

References

Aufderheide E, Ekblom P (1988) Tenascin in gut development: appearance in the mesenchyme, shift in molecular forms, and dependence on epithelial mesenchymal interactions. J Cell Biol 107:2341–2349
Aufderheide E, Chiquet-Ehrismann R, Ekblom P (1987) Epithelial-mesenchymal interactions in the developing kidney lead to expression of tenascin in the mesenchyme. J Cell Biol 105:599–608

Bonkhoff H, Wernert N, Dhom G, Remberger K (1991) Basement membranes in fetal, adult normal, hyperplastic and neoplastic human prostate. Virchows Arch [A] 418:375–381

Bonkhoff H, Wernert N, Dhom G, Remberger K (1992) Distribution of basement membranes in primary and metastatic carcinomas of the prostate. Hum Pathol 23:923–939

Castellucci M, Classen-Linke I, Mühlhauser J, Kaufmann P, Zardi L, Chiquet-Ehrismann R (1991) The human placenta: a model for tenascin expression. Histochemistry 95:449–458

Chiquet-Ehrismann R, Mackie EJ, Pearson CA, Sakakura T (1986) Tenascin: an extracellular matrix protein involved in tissue interactions during fetal development and oncogenensis. Cell 47:131–139

Chiquet-Ehrismann R, Kalla P, Pearson CA (1989) Participation of tenascin and transforming growth factor beta in reciprocal epithelial-mesenchymal interactions of MCF-7 cells and fibroblasts. Cancer Res 49:4322–4325

Crabb JW, Armes LG, Carr SA, Johnson CM, Roberts GD, Bordoli RS, McKeehan WL (1986a) Complete primary structure of prostatropin, a prostate epithelial cell growth factor. Biochemistry 25:4988–4993

Crabb JW, Armes LG, Johnson CM, McKeehan WL (1986b) Characterization of multiple forms of prostatropin (prostate epithelial growth factor) from bovine brain. Biochem Biophys Res Commun 136:1155–1161

Cunha GR, Donjacour A (1987) Stromal-epithelial interactions in normal and abnormal prostatic development. In: Coffey DS, Bruchovsky N, Gardner WA, Resnick MI, Karr JP (eds) Current concepts and approaches to the study of prostate cancer. Liss, New York, pp 251–272

Cunha GR, Donjacour AA, Cooke PS, Mee S, Bigsby RM, Higgins SJ, Sugimura J (1987) The endocrinology and developmental biology of the prostate. Endocr Rev 8:338–362

Dedhar S, Saulnier R, Nagle R, Overall CM (1993) Specific alterations in the expression of alpha 3 beta 1 and alpha 6 beta 4 integrins in the highly invasive and metastatic variants of human prostate carcinoma cells selected by in vitro invasion through reconstituted basement membrane. Clin Exp Metastasis 11:391–400

Ekblom P, Aufderheide E (1989) Stimulation of tenascin expression in mesenchyme by epithelial-mesenchymal interactions. Int J Dev Biol 33:71–79

Erickson HP, Bourdon MA (1989) Tenascin: an extracellular matrix protein prominent in specialized embryonic tissues and tumors. Annu Rev Cell Biol 5:71–92

Erickson HP, Lightner VA (1988) Hexabrachion protein (tenascin, cytotactin, brachionectin) in connective tissues, embryonic brain, and tumors. In: Miller KR (ed) Advances in cell biology, vol 2. JAI, London, pp 55–90

Fadok VA, Savill JS, Haslett C, Bratton DL, Doherty DE, Campbell PA, Henson PM (1992) Different populations of macrophages use either the vitronectin receptor or the posphatidylserine receptor to recognize and remove apototic cells. J Immunol 149:4029–4035

Ferguson JE, Schor AM, Howell A, Ferguson MW (1990) Tenascin distribution in the normal breast is altered during the menstrual cycle and in carcinomas. Differentiation 42:199–207

Fong CJ, Sherwood ER, Sutkowski DM, Abu-Jawdeh GM, Yokoo H, Bauer KD, Kozlowski JM, Lee C (1991) Reconstituted basement membrane promotes morphological and functional differentiation of primary human prostatic epithelial cells. Prostate 19:221–235

Frisch SM, Francis H (1994) Disruption of epithelial cell-matrix interactions induces apoptosis. J Cell Biol 124:619–626

Herlin M, Graeven U, Speicher D, Sela BA, Bennicelli JL, Kath R, DuPont Guerry IV (1991) Characterization of tenascin secreted by human melanoma cells. Cancer Res 51:4853–4858

Howeedy AA, Virtanen I, Laitinen L, Gould NS, Koukoulis GK, Gould VE (1990) Differential distribution of tenascin in the normal, hyperplastic, and neoplastic breast. Lab Invest 63:798–806

Ibrahim SN, Lightner VA, Ventimiglia JB, Ibrahim GK, Walther PJ, Bigner DD, Humphrey PA (1993) Tenascin expression in prostatic hyperplasia, intraepithelial neoplasia, and carcinoma. Hum Pathol 24:982–989

Inaguma Y, Kusakabe M, Mackie, EJ, Pearson CA, Chiquet-Ehrismann R, Sakakura T (1988) Epithelial induction of stromal tenascin in mouse mammary gland: from embryogenesis to carcinogenesis. Dev Biol 128:245–255

Koch M, Wehrle-Haller B, Baumgartner S, Spring J, Brubacher D, Chiquet M (1991) Epithelial synthesis of tenascin at tips of growing bronchi and graded accumulation in basement membrane and mesenchyme. Exp Cell Res 194:297–300

Koukoulis GK, Virtanen I, Korhonen M, Laitinen L, Quaranta V, Gould VE (1991) Immunohistochemical localization of integrins in the normal, hyperplastic and neoplastic breast. Am J Pathol 139:787–799

Lightner VA, Gumkowski F, Bigner DD, Erickson HP (1989) Tenascin/hexabrachion in human skin: biochemical identification and localization by light and electron microscopy. J Cell Biol 108:2483–2493

Lightner VA, Marks JR, McCachren SS (1994) Epithelial cells are an important source of tenascin in normal and malignant human breast tissue. Exp Cell Res 210:177–184

Mackie EJ, Halfter W, Liverani D (1988) Induction of tenascin in healing wounds. J Cell Biol 107:2757–2767

Mackie EJ, Scott-Burdon T, Hahn AW, Kern F, Bernhardt J, Regenas S. Weller A, Bühler FR (1992) Expression of tenascin by vascular smooth muscle cells. Alterations in hypertensive rats and stimulation by angiotensin II. Am J Pathol 141:377–388

McCachren SS, Lightner VA (1992) Expression of tenascin in synovitis and its regulation by interleukin-1. Arthritis Rheum 35:1185–1196

McKeehan WL (1991) Growth factor receptors and prostate cell growth. In: Isaacs JT, Franks LM (eds) Prostate cancer: cell and molecular mechanisms in diagnosis and treatment. Cold Spring Harbor Press, Cold Spring Harbor, pp 165–175

Meredith JE Jr, Fazeli B, Schwartz MA (1993) The extracellular matrix as a survival factor. Mol Biol Cell 4:953–961

Passaniti A, Isaacs JT, Haney JA, Adler SW, Cujdik TJ, Long PV, Kleinman HK (1992) Stimulation of human prostatic carcinoma tumor growth in athymic nude mice and control of migration in culture by extracellular matrix. Int J Cancer 51:318–324

Pearson CA, Pearson D, Shibahara S, Hofsteenge J, Chiquet-Ehrismann R (1988) Tenascin: cDNA cloning and induction by TGF-bega. EMBO J 7:2977–2982

Ramadori G, Schwögler S, Veit T, Rieder H, Chiquet-Ehrismann R, Mackie EJ, Meyer zum Büschenfelde K-H (1991) Tenascin gene expression in rat liver and rat liver cells. In vivo and in vitro studies. Virchows Arch [B] 60:145–153

Reed J (1994) Bcl-2 and the regulation of programmed cell death. J Cell Biol 124:1–6

Reid DM, Perry VH, Andersson PB, Gordon S (1993) Mitosis and apoptosis of microglia in vivo induced by an anti-CR3 antibody which crosses the blood-brain barrier. Neuroscience 56:529–533

Ruoslahti E, Reed JC (1994) Anchorage dependence, integrins, and apoptosis. Cell 77:477–478

Savill J, Dransfield I, Hogg N, Haslett C (1990) Vitronectin receptor-mediated phagocytosis of cells undergoing apoptosis. Nature 343:170–173

Schalwijk J, Steijlen PM, van Vlijmen-Willems IMJJ, Oosterling B, Mackie EJ, Verstaeten AA (1991) Tenascin expression in human dermis is related to epidermal proliferation. Am J Pathol 139:1143–1150

Schwogler S, Odenthal M, Meyer zum Burschenfelde KH, Ramadori G (1992) Alternative splicing products from arterial smooth muscle cells and skin fibroblasts. Biochem Biophys Res Commun

Story MT (1991) Polypeptide modulators of prostatic growth and development. In: Isaacs JT, Franks LM (eds) Prostate cancer: cell and molecular mechanisms in diagnosis and treatment. Cold Spring Harbor Press, Cold Spring Harbor, pp 123–146

Tenniswood MP, Guenette RS, Lakins J, Mooibroek M, Wong P, Welsh J-E (1992) Active cell death in hormone-dependent tissues. Cancer Metastasis Rev 11:197–220

Tucker RP, Hammarback JA, Jenrath DA, Mackie EJ, Xu Y (1993) Tenascin expression in the mouse: in situ localization and induction by bFGF. J Cell Sci 104:69–76

Vollmer G, Siegal GP, Chiquet-Ehrismann R, Lightner VA, Arnholdt H, Knuppen R (1990) Tenascin expression in the human endometrium and in endometrial adenocarcinomas. Lab Invest 62:725–730

Vollmer G, Deerberg F, Siegal GP, Knuppen R (1991) Altered tenascin expression during spontaneous endometrial carcinogenesis in the BDII/Han rat. Virchows Arch [B] 60:83–89

Vollmer G, Michna H, Ebert K, Knuppen R (1992) Down-regulation of tenascin expression by antiprogestins during terminal differentiation of rat mammary tumors. Cancer Res 52:4642–4648

Washizu K, Kimura S, Hiraiwa H, Matsunaga K, Kuwabara M, Ariyoshi Y, Kato K, Takeuchi K (1993) Development and application of an enzyme immunoassay for tenascin. Clin Chim Acta 219:15–22

Wilding G (1991) Response of prostate cancer cells to peptide growth factors: transforming growth factor-B. In: Isaacs JT, Franks LM (eds) Prostate cancer: cell and molecular mechanisms in diagnosis and treatment. Cold Spring Harbor Press, Cold Spring Harbor, pp 147-163

Yavin E, Gabai A, Gil S (1991) Nerve growth factor mediates monosialoganglioside induced release of fibronectin and J1/tenascin. J Neurochem 56:105–112

8 Anti-Growth Factor Activity of Antiestrogens in Human Breast Cancer Cells: A Review

F. Vignon and H. Rochefort

8.1 Introduction

The conjunction of the long-recognized role of ovarian hormones on breast tumor growth (Beatson 1896) and the identification of intracellular estrogen receptors (ER; Jensen and Jacobson 1960) has prompted the design and therapeutic use of drugs which could selectively antagonize unfavorable steroid hormone action in their target tissues. Nonsteroidal antiestrogens were initially described by Lerner and coworkers in 1958; many structural derivatives of triphenylethylene have since been synthesized, the most widely used in breast cancer treatment being tamox-

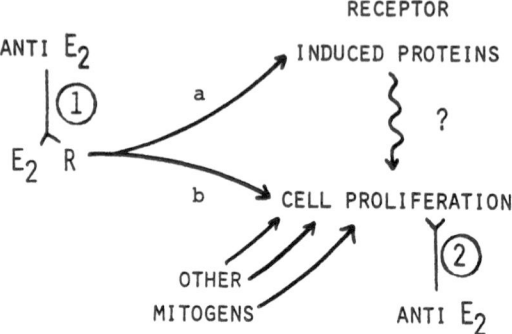

Fig. 1. Antiestrogen action on the two series of estrogen-regulated responses. Estradiol (E_2) via its receptor (R) specifically stimulates the synthesis of some proteins (a). Antiestrogen (*Anti E_2*) inhibits this effect by competing on the R and inducing a defective R activation (Step *1*). The mechanism of action of Anti E_2 for inhibiting cell growth may be more complex since besides E_2, other mitogens stimulate cell proliferation (b). Does Anti E_2 simply inhibit E_2 effect (antiestrogenic effect) (Step *1*), or does it also prevent cell proliferation by other mechanisms (anti proliferative effect; Step *2*)? (From Rochefort et al. 1983)

ifen (or Nolvadex) (Harper and Walpole 1966). Though their pharmacology is complex and often paradoxical, they all present common characteristic features. They interact with nuclear ER and display estrogen agonist activities which vary within species (human, chick, mouse, rat), tissue (uterus, breast, bone), or with the gene considered (progesterone receptor, cathepsin D, pS2). In the presence of estrogens, they act as competitive inhibitors for binding to nuclear steroid receptors and thus behave as strong hormone antagonists. On the basis of these recognized antagonistic properties, several antiestrogens have been tested clinically as agents for breast cancer therapy (Cole et al. 1971).

Current interest in the antiestrogens is then focused upon the development of compounds with a higher affinity for the ER with an anticipated result of increased potency. In fact, hydroxylation of some antiestrogens dramatically increases their affinity (Jordan et al. 1977; Rochefort et al. 1979); however, this advantage might be offset by a decrease in the biological half-life of the hydroxylated molecule. A number of steroidal and nonsteroidal compounds devoid of any agonist activity have emerged and now await clinical validation (Wakeling and Bowler 1987, 1992; Van de Velde et al. 1994).

Antiestrogens have become commonly used drugs in the treatment of breast cancer although the mechanisms by which they evoke their antagonistic activity remain poorly understood. The availability of several hormone-responsive and unresponsive human breast cancer cell lines (Lippman et al. 1976) has allowed the development of molecular and cellular studies on in vitro model systems as well as on their in vivo counterparts as tumors grown in athymic nude mice (Soule and McGrath 1980).

The number of cancer cells can be reduced by antiestrogens according to different mechanisms. Antiestrogens can decrease cell proliferation via their antiestrogenic and antigrowth factor activity; they can also increase cell death through a mechanism which has been studied much less. In all cases, however, they appear to be mostly active via the ER located in cancer cells (Fig. 1) and to behave as specific ER-targeted compounds (Rochefort 1987). We will successively consider these different aspects of antiestrogen action in breast cancer cell.

8.2 Antiestrogenic Activities of Antiestrogens

The clinical efficacy of nonsteroidal antiestrogens in therapy of breast cancer patients with ER+ tumors as well as in vivo and in vitro experimental data have shown the key role of ER in mediating antiestrogen inhibitory action. Studies with cultured human breast cancer cells which differ in their ER levels demonstrate that antiestrogens selectively inhibit the proliferation of ER+ cells (MCF7, T47D, or ZR 75–1) while the growth of ER– cell lines (MDA MB231, BT20) is unaffected (Lippman et al. 1976). The sensitivity of different breast tumor cell lines to various antiestrogens correlates well with their nuclear ER levels and with the relative affinity for ER of the tested compounds (Coezy et al. 1982). Moreover, the suppression of cell growth by antiestrogens is reversed by estradiol, suggesting that antiestrogens exert growth-suppressive effects by competing with estrogens for binding to their specific receptors within the tumor cells in the presence of estrogen agonist (steroid hormone or phenol red contaminants; Berthois et al. 1986).

In addition to their stimulatory effect on tumor growth, estrogens regulate the expression of several genes, some of which have been associated with secreted growth factors and mitogens (Westley and Rochefort 1980; Dickson and Lippman 1987). These results led us to propose the hypothesis of an autocrine loop partly or totally mediating the mitogenic activity of estrogens in breast cancer (Rochefort et al. 1980; Vignon et al. 1986). Antiestrogens, acting as competitors of estrogens, not only neutralize direct estrogen action but also strongly decrease the expression of estrogen-induced mitogens. Moreover, anti-estrogens were proposed to increase the secretion of active growth inhibitors, such as transforming growth factor-β_1 (TGF-β_1) in some hormone-responsive cells (Knabbe et al. 1987). The antiestrogenic properties of these antagonists could thus result from the conjunction of regulatory actions on estrogen agonists and on estrogen-induced positive and negative responses.

The growth of human breast tumor cell lines is coordinately promoted by estrogens, peptide hormones (insulin), and growth factors acting via transmembrane receptors (insulin-like growth factor, IGF-I, IGF-II; epidermal growth factor, EGF; TGF-α; or other proteins) (for reviews, see Dickson and Lippman 1987; Vignon and Rochefort 1987). They represent a useful tool in evaluating whether steroid antagonists act exclusively as estrogen antagonists or whether they may prevent tumor proliferation by additional antiproliferative activities.

8.3 Anti-Growth Factor Activity of Antiestrogens

The fact that in MCF7 cells antiestrogens inhibited cell proliferation in the apparent absence of estrogens and that some of the growth factors were very active in stimulating the growth of these cells suggested that these steroid antagonists could also inhibit growth factor action (Rochefort et al. 1983; Fig. 1).

When stimulating MCF7 cells with insulin, IGF-I, or EGF, in estrogen-free conditions, we in fact showed that the nonsteroidal antiestrogen 4-hydroxy tamoxifen could drastically inhibit mitogen-induced cell proliferation (Vignon et al. 1987; Freiss et al. 1993). This non-antiestrogenic effect of tamoxifen is also mediated by ERs: it is not observed in ER– cells; the relative efficacy of Tam and its hydroxylated

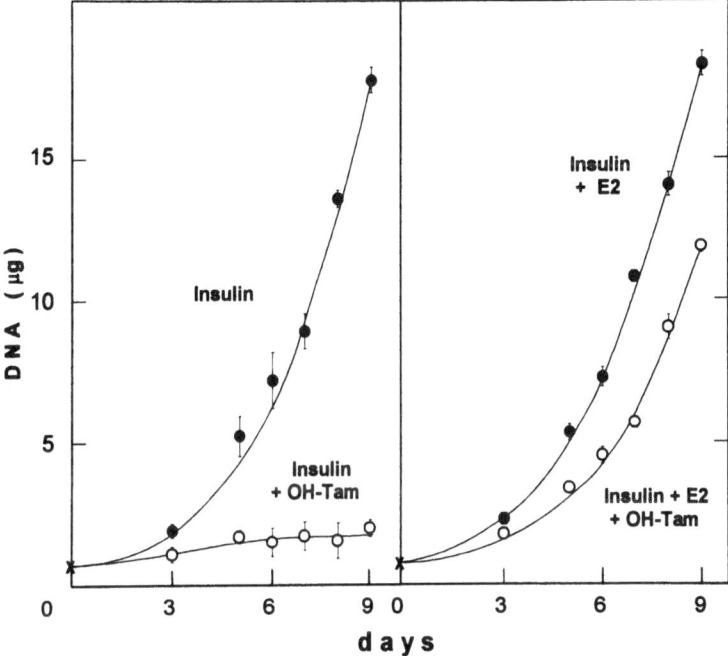

Fig. 2. Anti-insulin activity of OH-Tam. MCF7 cells, maintained in charcoal-stripped serum for 7 days, were grown for increasing periods of time under four different treatment conditions: *left panel,* insulin (50 n*M*; *solid circles*) or insulin (50 n*M*) and OH-Tam (10 n*M*; *open circles*) added simultaneously; *right panel,* insulin (50 n*M*) and estradiol (E$_2$; 100 n*M*; *solid circles*) or insulin (50 n*M*), E$_2$ (100 n*M*) and OH-Tam (10 n*M; open circles*). Each *point of the curves* represent the mean ± SD of three separate DNA determinations performed with the diaminobenzoic acid (DABA) assay. (Redrawn from Vignon et al. 1987)

metabolite is correlated with their affinity for ER and is selectively rescued by addition of estradiol (Fig. 2).

These results thus demonstrated that nonsteroidal antiestrogens can inhibit growth not only by blocking estrogen action but also by strongly impeding growth factor mitogenic activities.

Our initial observation was confirmed in various in vitro studies on breast cancer cells (Berthois et al. 1989; Wakeling et al. 1989) and more

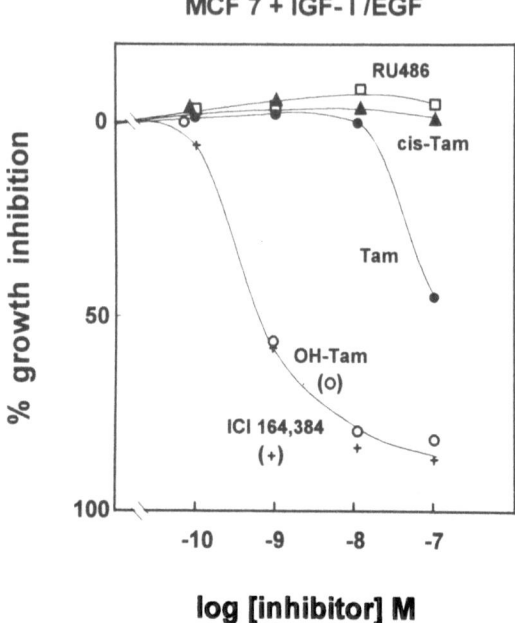

Fig. 3. Comparative anti-growth factor activity of steroidal or nonsteroidal antiestrogens and of an antiprogestin (RU486). MCF7 cells, withdrawn from steroids, were grown for 7 days in the presence of insulin-like growth factor-I (IGF-I; 5 n*M*) or epidermal growth factor (EGF; 4 n*M*) together with increasing concentrations of ICI 164,384, OH-Tam, Tam, *cis*-Tam and RU486. The percentage of growth inhibition versus EGF-stimulated or IGF-I-stimulated cells was calculated after DNA determinations on triplicate wells for each treatment condition. (Redrawn from Freiss et al. 1990a)

recently in vivo in an experiment on mouse uterus (Ignar-Trowbridge et al. 1992). This anti-growth factor activity is not restricted to a class of estrogen antagonists since pure steroidal antiestrogens (ICI 164,384 or ICI 182,780) also display an inhibitory effect (Freiss et al. 1990a). Moreover, other steroid nuclear receptor ligands such as retinoic acid or glucocorticoid can also exert a strong antiproliferative activity in breast tumors (Freiss et al. 1990a) or other cellular systems. However, this anti-growth factor activity is not a common feature of all ligands of steroid receptor superfamily and can be clearly dissociated from antiestrogenic activity since synthetic progestin (R5020) (Vignon et al. 1982) or antiprogestin (RU 38,486) (Bardon et al. 1985), which both efficiently antagonize estradiol growth-promoting activity (Gill et al. 1987), are totally ineffective in the sole presence of growth factors (Freiss et al. 1990a; Poulin et al. 1989; Murphy and Dotzlaw 1989) (Fig. 3).

Finally, this inhibitory effect of OH-Tam not only affects the mitogenic response but also some other growth factor-induced responses: (a) the regulation of the progesterone receptor protein, whose increased synthesis following EGF or IGF-I treatment is totally depleted by antiestrogen (Sumida and Pasqualini 1989; Katzenellenbogen and Norman 1990), and (b) the EGF-stimulated or IGF-I-stimulated transcriptional activity of several genes including pS2 and cathepsin D, which is totally abrogated by steroidal or nonsteroidal antagonists (Chalbos et al. 1993).

To further analyze the mechanisms by which steroid nuclear receptor ligands could prevent growth factor action, we have analyzed how estrogen antagonists can impede, at the membrane level, the initial steps of EGF or IGF-I action or, at the nuclear level, the regulation and activity of immediate-early gene products.

8.4 Membrane and Nuclear Mechanisms of Anti-Growth Factor Activity

8.4.1 Membrane Cross Talk

At the plasma membrane, the first steps of growth factor action include binding to the extracellular domains of their transmembrane receptors and rapid activation of their intrinsic tyrosine kinase activity, which leads to receptor autophosphorylation on tyrosine residues, now recognized as a key event for recognition of intracellular second messengers (Pazin and Williams 1992).

After excluding the possibility that OH-Tam could inhibit growth factor binding by direct competition on their respective binding sites, we have studied how estrogens and their antagonists could modulate growth factor binding parameters (concentrations of sites, affinity; Freiss et al. 1990b). In ER+ MCF7 breast cancer cell line, we have showed that OH-Tam had an opposite effect on EGF and IGF-I binding.

Whereas IGF-I binding decreased by 60% after 6 days of antagonist treatment, high-affinity EGF binding was tripled within the same period. The affinities of EGF and IGF-I for their respective receptors were neither modified by estrogen nor by antagonist. These results were extended to another ER+ breast cancer cell line (T47D); by contrast, in ER– BT20 cells, EGF and IGF-I binding were not affected by estradiol or its antagonist, indicating that these regulatory effects required the presence of ER.

The parallel between the decrease in IGF-I mitogenic activity and the decrease in IGF-I binding strongly suggests that one mechanism by which OH-Tam could impede growth factor action is by exerting a negative regulation of their binding sites. On the contrary, the contrasting results on EGF binding and EGF mitogenic activity rather favor the presence of an alternative regulatory mechanism such as an impaired transducing system. Since EGF receptor autophosphorylation on tyrosine residues is essential for mitogenic signal transduction, we have carried out experiments to evaluate whether OH-Tam could affect the state of EGF receptor tyrosine phosphorylation. We have shown that OH-Tam decreases phosphorylation of EGF receptor in MCF 7 cells by 90% (Freiss et al. 1990b) whereas it was inefficient in modulating EGF receptor phosphorylation in two ER– cells. Thus nonsteroidal antagonists can also affect growth factor signal transduction.

The mechanism by which EGF receptor autophosphorylation is decreased (decrease in tyrosine kinase activity or increase in protein tyrosine phosphatase (PTPase) activity) is under current investigation. Our present results suggest that steroidal and nonsteroidal antiestrogens can regulate PTPase activity in ER+ human breast cancer cells (Freiss and Vignon 1994) though the enzyme identity and its mechanism of regulation remain unknown.

8.4.2 Nuclear Transcriptional Cross Talk

Increase in AP-1 activity is one of the first nuclear event triggered by most growth factors. We have have thus evaluated in our laboratory whether estrogens and antiestrogens can affect this growth factor-induced activity in conditions where their mitogenic action could be modulated. Using transient transfection of an AP-1 responsive gene

[(AP-1)$_4$-TK-CAT)] in ER+ MCF7 cells, our laboratory showed that estradiol increased AP-1 activity whereas antiestrogens (OH-Tam, ICI 164,384) were strongly inhibitory (Philips et al. 1993). This inhibition occurred without any modulation of the levels of c-*fos* and c-*jun* mRNAs and in conditions in which basal ERE-mediated activity levels are unchanged. ER ligands such as OH-Tam and ICI 164,384 are thus able to modulate not only ERE-mediated trancription but also growth factor-regulated AP-1 activity.

Our current molecular knowledge on antiestrogen action in human breast cancer cells indicates that steroidal and nonsteroidal compounds can inhibit their proliferation by strongly antagonizing both steroid and growth factor action. Since these drugs are also active in the absence of estrogens, they should be better named *"ER-mediated growth inhibitors"* than antiestrogens (Rochefort 1987). Independently of their numerous systemic effects on the host target tissues, administration of antiestrogens results in the mammary tumor in a strong blockade of estrogen and/or growth factor activities which will drastically affect the tumor cellular proliferative program.

8.5 Antiestrogens and Active Cell Death

Decrease of human breast cancer cell population in vitro or tumor regression in vivo following antiestrogen treatment reflects a change in the balance of cellular growth events and can involve an arrested cell proliferation or an enhanced cell death or both. Several studies were thus initiated with the different classes of antiestrogens to evaluate their impacts on cell cycle kinetics and parameters to ascertain whether these compounds act primarily as cytostatic and/or as cytotoxic drugs. While all data agree on the fact that these drugs present a strong cytostatic activity, their influence on cell death has long been controversial.

8.5.1 Cytostatic Activities of Antiestrogens

Green et al. (1981) first showed, employing the technique of analytical DNA flow cytometry, that the tamoxifen-induced decrease in breast cancer cell proliferation rate is associated with an accumulation of cells in the G_0/G_1 phase of the cell cycle. This observation, later confirmed by others (Osborne et al. 1983; Benz et al. 1983), was applicable to asynchronous as well as synchronous cells. Tamoxifen treatment results in a dose-dependent reduction in the S and G_2/M fractions and a concomitant increase in the G_1 fraction in which more than 90% of cells are accumulated after 96 h. When the S phase is estimated autoradiographically by the [3H]thymidine labeling index (TLI), tamoxifen results in a drastic decrease in the proportion of cells actively engaged in DNA synthesis (5%). Similar effects are observed when treating cells with other antiestrogens (nafoxidine, trioxifene) thus confirming that antiestrogens slow breast cancer cell proliferation by initiating a block in early G_1 phase (Osborne et al. 1983). Antiestrogen inhibition in vitro can be reversed by addition of estradiol, suggesting that short-term drug exposure is not lethal.

Similarly, some in vivo studies on MCF7 tumors in athymic nude mice concluded that both antiestrogen administration and estrogen withdrawal induced a stationary phase of tumor growth without any histological sign of cell death (Osborne et al. 1985; Gottardis et al. 1988). The facts that estrogen replenishment restores tumor growth and that tumor cell viability is maintained despite prolonged tamoxifen treatment argued in favor of a prevalent cytostatic effect. However, other in vivo studies showed some tumor regression after estrogen deprivation and tamoxifen treatment (Shafie and Grantham 1981) and therefore suggested that these compounds can also increase cell death.

8.5.2 Antiestrogens and Apoptosis

The first evidence of "ER-mediated cytotoxicity" following hydroxylated tamoxifen treatment originated from a study performed in our laboratory by Bardon et al. (1987). In this study performed on human breast cancer cells, we showed that treatment with nanomolar concentrations of antiestrogens (and antiprogestins) produces cytotoxic and cytostatic effects. Several biochemical and morphological parameters of cell death were evaluated with a particular emphasis on the search for apoptotis, a process of active cell death (Kerr et al. 1972).

This process, now known to be associated with specific gene expression or protein synthesis (for review, Tenniswood et al. 1992), has been initially defined by its characteristic morphological modifications. Apoptosis asynchronously affects a cell population and involves a disruption of cell–cell and cell–basement membranes, connections contributing to isolate the cell which undergoes active cell death from its neighbors. As a consequence of active DNA hydrolysis by endonucleases, chromatin condensation occurs and produces hyperchromatic, piknotic nuclei (Kerr et al. 1972). At a more advanced stage, apoptotic cellular bodies containing membrane-bound portions of fragmented nuclei as well as intact cytoplasmic organelles and lysosomes are released in the extracellular space. These apoptotic bodies, which export cellular material without leakage, by contrast to what is observed in necrosis, are further phagocytosed and degraded by surrounding cells and macrophages.

In antiestrogen-treated MCF7 cells, the number of dead cells, estimated by a dye-exclusion technique, was doubled within a few days and represented two thirds of the total cell population at the end of treatment (Bardon et al. 1987). Estradiol, at concentrations sufficient to displace OH-Tam from the ER, prevented cell death and maintained a high rate of cell mitosis. No such antiestrogen-induced cell death could be detected in ER− human breast cancer cells (MDA MB231, BT20). Moreover, the increase of unsedimented DNA in antiestrogen-treated cells suggested that active DNA fragmentation into nucleosomal oligomers had occurred. Antiestrogens (and antiprogestins) were shown to induce a detectable active cell death process in ER+ cells, even though the presence of phenol red, identified as an estrogen contaminant (Berthois et al. 1986), might have weakened the phenomenon somewhat (Bardon et al. 1987). An ongoing study in our laboratory (P. Roger, unpublished observations) has confirmed, by the in situ nick-end labeling technique (Gavrieli et al. 1992; Wijsman et al. 1993) that OH-Tam and a pure antagonist (ICI 164,384) are able to induce apoptosis in cultured MCF7 cells. With the terminal-transferase assay, 3' end DNA breaks indicative of DNA fragmentation can be detected histochemically on frozen or paraffin-embedded sections of human tumor biopsies and have proven to be adaptable to estimate whether apoptosis occurs in vivo within tumors of tamoxifen-treated patients (P. Roger, T. Maudelonde, F. Vignon and H. Rochefort, unpublished data).

Moreover, two recent studies on human MCF7 cells growing as tumors in athymic nude mice have indicated that apoptosis occurred in vivo following estrogen withdrawal (Kyprianou et al. 1991) or antiestrogen treatment (Wärri et al. 1993). Tumor regression was characterized by a drastic reduction in the mitotic index and a simultaneous increase in apoptotic index (30%–60% of cells undergoing apoptosis). The important morphological changes were accompanied by rapid and transient expression of c-*myc* (Kyprianou et al. 1991) and TRPM-2 and TGF-β1, two genes previously associated to apoptosis in the involuting prostate (Tenniswood et al. 1992; Kyprianou and Isaacs 1989). The results of both groups agreed on the timing of increase of these genes (optimal increase at day 3) and on the restrictive specificity of their induction since estrogen-responsive genes (pS2) or proliferation-associated genes (c-*fos* and c-H-*ras*) were rapidly downregulated in these regressing tumors. However, the characteristic ladder of DNA nucleosomal oligomers was not detected in toremifene-treated cells while appearing as early as 1 day after estradiol ablation in the other experimental design. Whether this reflects a true difference in apoptotic processes following different modes of hormonal manipulation of tumor growth or is due to the proportion of cells undergoing apoptosis remains unknown.

8.6 Conclusions – Prospects

Several studies have shown, by different techniques, that active cell death, which naturally occurs in mammary gland involution after weaning, also contributes to tumor regression following estrogen ablation or antiestrogen treatment. However, contrary to the drastic effect observed with glucocorticoids on thymocytes (Wyllie 1980), fewer cells undergo apoptosis after treatment of breast cancer cells with antiestrogens or antiprogestins. The demonstration that these drugs could strongly antagonize both steroid and growth factor-induced mitogenic and transcriptional activities and that their administration could thus lead to a deprivation of both steroids and growth factors within tumors strengthens the idea that these antagonists could act as apoptotic agents (Fig. 4).

An increasing number of genes (*Bcl2*, p53, TRPM-2, ICE, *ced-3*; see other chapters, this volume) have been associated to the onset of apoptosis (Vaux et al. 1994; Tenniswood et al., this volume). For breast cancer, future studies will be aimed at understanding the mechanism of receptor-in-

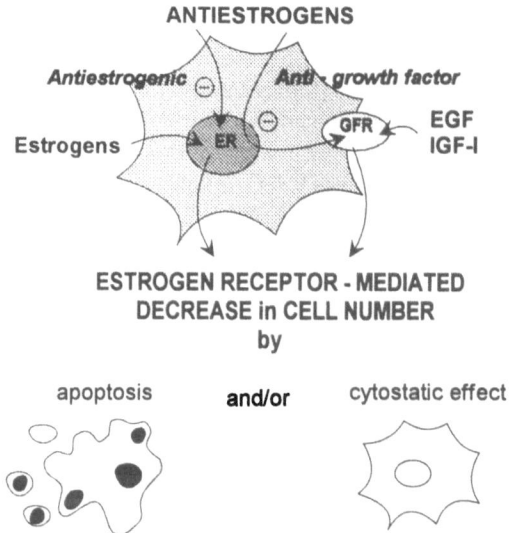

Fig. 4. Control of breast cancer cell growth by antiestrogens: 1994 schematic representation. In addition to the well-documented cytostatic effect, antiestrogens might also induce apoptosis by blocking steroid and growth factor action in breast cancer cells. *IGF-I,* insulin-like growth factor-I; *EGF,* epidermal growth factor; *ER,* estrogen receptor; *GFR,* growth factor receptor

duced active cell death and searching how the combining effects of steroidal or nonsteroidal antiestrogens with other associated cancer treatments can further increase apoptosis. A new therapeutic strategy of breast cancer, which is currently focused on the blockade of proliferation and invasion, might be to induce selective stimulation of apoptosis.

Acknowledgments. We are grateful to N. Kerdjadj for her help in the preparation of the manuscript. This work was supported by funding from the Institut National de la Santé et de la Recherche Médicale, the University of Montpellier I, the Association pour la Recherche sur le Cancer (grants 6737, 6757, 1250), the Ligue Nationale contre le Cancer and the Fédération Nationale des Centres de lutte contre le Cancer (grant 257 705 097).

References

Bardon S, Vignon F, Chalbos D, Rochefort H (1985) RU486, a progestin and glucocorticoid antagonist inhibits the growth of breast cancer cells via the progesterone receptor. J Clin Endocrinol Metab 60:692–697

Bardon S, Vignon F, Montcourrier P, Rochefort H (1987) Steroid-receptor mediated cytotoxicity of an antiestrogen and an antiprogestin in breast cancer cells. Cancer Res 47:1441–1448

Beatson GT (1896) On the treatment of inoperable cases of carcinoma of the mamma: suggestions for a new method of treatment, with illustrative cases. Lancet 2:104–107, 162–167

Benz C, Cadman E, Gwin J, Wu T, Amara J, Eisenfeld A, Dannies P (1983) Tamoxifen and 5-fluorouracil in breast cancer: cytotoxic synergism in vitro. Cancer Res 43:5298–5303

Berthois Y, Katzenellenbogen JA, Katzenellenbogen BS (1986) Phenol red in tissue culture media is a weak estrogen: implications concerning the study of estrogen-responsive cells in culture. Proc Natl Acad Sci USA 83:2496–2500

Berthois Y, Dong XF, Martin PM (1989) Regulation of epidermal growth factor receptor by estrogen and antiestrogen in the human breast cancer cell line MCF-7. Biochem Biophys Res Commun 159:126–131

Chalbos D, Philips A, Galtier F, Rochefort H (1993) Synthetic antiestrogens modulate induction of pS2 and cathepsin D mRNA by growth factors and adenosine 3', 5'-monophosphate in MCF7 cells. Endocrinology 133:571–576

Coezy E, Borgna JL, Rochefort H (1982) Tamoxifen and metabolites in MCF7 cells: correlation between binding to estrogen receptor and inhibition of cell growth. Cancer Res 42:317–323

Cole MP, Jones CTA, Todd IDH (1971) A new anti-estrogenic agent in late breast cancer. Br J Cancer 25:270–275

Dickson RB, Lippman ME (1987) Estrogenic regulation of growth and polypeptide growth factor secretion in human breast carcinoma. Endocrine Rev 8:29–43

Freiss G, Vignon F (1994) Antiestrogens increase protein tyrosine phosphatase activity in human breast cancer cells. Mol Endocrinol 8:1389–1396

Freiss G, Prébois C, Rochefort H, Vignon F (1990a) Anti-steroidal and anti-growth factor activity of antiestrogens. J Steroid Biochem Mol Biol 37:777–781

Freiss G, Rochefort H, Vignon F (1990b) Mechanisms of 4-hydroxytamoxifen anti-growth factor activity in breast cancer cells: alterations of growth factor receptor binding sites and tyrosine kinase activity. Biochem Biophys Res Commun 173:919–926

Freiss G, Prébois C, Vignon F (1993) Control of breast cancer cell growth by steroids and growth factors: interactions and mechanisms. Breast Cancer Res Treat 27:57–68

Gavrieli Y, Sherman Y, Ben-Sasson SA (1992) Identification of programmed cell death in situ via specific labeling of nuclear DNA fragmentation. J Cell Biol 119:493–501

Gill PG, Vignon F, Bardon S, Derocq D, Rochefort H (1987) Difference between R5020 and the antiprogestin RU486 in antiproliferative effects on human breast cancer cells. Breast Cancer Res Treat 10:37–45

Gottardis MM, Robinson SP, Jordan VC (1988) Estradiol-stimulated growth of MCF-7 tumors in athymic nude mice: a model to study the tumoristic action of tamoxifen. J Steroid Biochem 20:311–314

Green MD, Whybourne AM, Taylor IW, Sutherland RL (1981) Effects of antioestrogens on the growth and cell cycle kinetics of cultured human mammary carcinoma cells. In: Sutherland RL, Jordan VC (eds) Non-steroidal antioestrogens: molecular pharmacology and antitumour activity. Academic, Sydney, pp 397-412

Harper MJK, Walpole AL (1966) Contrasting endocrine activities of cis and trans isomers in a series of substituted triphenylethylenes. Nature 212:87

Ignar-Trowbridge DM, Nelson KG, Bidwell MC, Curtis SW, Washburn TF, Mc Lahlan JA, Korach KS (1992) Coupling of dual signaling pathways: epidermal growth factor action involves the estrogen receptor. Proc Natl Acad Sci USA 89:4658–4662

Jensen EV, Jacobson HI (1960) Fate of steroid estrogens in target tissues. In: Pincus G, Vollmer EP (eds) Biological activity of steroids in relation to cancer. Academic, New York, pp 161–178

Jordan VC, Collins MM, Rowsby L, Prestwich G (1977) A monohydroxylated metabolite of tamoxifen with potent antioestrogenic activity. J Endocrinol 75:305–316

Katzenellenbogen BS, Norman MJ (1990) Multihormonal regulation of the progesterone receptor in MCF7 human breast cancer cells: interrelationships among insulin/insulin-like growth factor-I, serum and estrogen. Endocrinology 126:891–898

Kerr JFK, Wyllie AH, Currie AH (1972) Apoptosis, a basic biological phenomenon with wider implications in tissue kinetics. Br J Cancer 26:239–245

Knabbe CK, Lippman ME, Wakefield LM, Flanders KC, Kasid A, Derynck R, Dickson RB (1987) Evidence that transforming growth factor-beta is a hormonally regulated negative growth factor in human breast cancer cells. Cell 48:417–428

Kyprianou N, Isaacs JT (1989) Expression of transforming growth factor-β in the ventral prostate during castration-induced programmed cell death. Mol Endocrinol 3:1515–1522

Kyprianou N, English HF, Davidson NE, Isaacs JT (1991) Programmed cell death during regression of the MCF-7 human breast cancer following estrogen ablation. Cancer Res 51:162–166

Lerner LJ, Holthaus JF, Thompson CR (1958) A non-steroidal estrogen antagonist 1-(p-2-diethylaminoethoxyphenyl)-1-phenyl-2-p-methoxyphenylethanol. Endocrinology 63:295–318

Lippman ME, Bolan G, Huff K (1976) The effects of estrogens and antiestrogens on hormone-responsive human breast cancer in long term culture. Cancer Res 36:4595–4601

Murphy LC, Dotzlaw H (1989) Endogenous growth factor expression in T47D human breast cancer cells associated with reduced sensitivity of antiproliferative effects of progestins and antiestrogens. Cancer Res 49:599–604

Osborne CK, Boldt DH, Clark GM, Trent JM (1983) Effects of tamoxifen on human breast cancer cell kinetics: accumulation of cells in early G_1 phase. Cancer Res 43:3583–3585

Osborne CK, Hobbs K, Clark GM (1985) Effect of estrogens and antiestrogens on growth of human breast cancer cells in athymic nude mice. Cancer Res 45:584–590

Pazin MJ, Williams LT (1992) Triggering signaling cascades by receptor tyrosine kinases. Trends Biochem Sci 17:374–375

Philips A, Chalbos C, Rochefort H (1993) Estradiol increases and anti-estrogens antagonize the growth factor-induced AP-1 activity in MCF7 breast cancer cells without affecting c-*fos* and c-*jun* synthesis. J Biol Chem 268:14103–14108

Poulin R, Dufour JM, Labrie F (1989) Progestin inhibition of estrogen-dependent proliferation in ZR 75-1 human breast cancer cells: antagonism by insulin. Breast Cancer Res Treat 13:265–276

Rochefort H (1987) Do antiestrogens and antiprogestins act as hormone antagonists or receptor-targeted drugs in breast cancer ? Trends Pharmacol Sci 8:126–128

Rochefort H, Garcia M, Borgna JL (1979) Absence of correlation between antiestrogenic activity and binding affinity for the estrogen receptor. Biochem Biophys Res Commun 88:351–357

Rochefort H, Coezy E, Joly E, Westley B, Vignon F (1980) Hormonal control of breast cancer in cell culture. In: Iacobelli S, King RJB, Lindner HR, Lippman ME (eds) Hormones and cancer, progress in cancer research and therapy, vol 14. Raven, New York, pp 21–29

Rochefort H, Borgna JL, Evans R (1983) Cellular and molecular mechanism of action of antiestrogens. J Steroid Biochem 19:69–74

Shafie SM, Grantham FH (1981) Role of hormones in the growth and regression of human breast cancer cells (MCF-7) transplanted into athymic nude mice. J Natl Cancer Inst 67:51–56

Soule HD, Mc Grath CM (1980) Estrogen responsive proliferation of clonal human breast carcinoma cells in athymic nude mice. Cancer Lett 10:177–189

Sumida C, Pasqualini JR (1989) Antiestrogens antagonize the stimulatory effect of epidermal growth factor on the induction of progesterone receptor in fetal uterine cells in culture. Endocrinology 124:591–597

Sutherland RL, Hall RE, Taylor IW (1983) Cell proliferation kinetics of MCF7 human mammary carcinoma cells in culture and effects of tamoxifen on exponentially growing and plateau-phase cells. Cancer Res 43:3998–4006

Tenniswood MP, Guenette RS, Lakins J, Mooibroek M, Wong P, Welsh JE (1992) Active cell death in hormone-dependent tissues. Cancer Metastasis Rev 11:192–220

Van de Velde P, Nique F, Bouchoux F, Brémaud J, Hameau MC, Lucas D, Moratille C, Viet S, Philibert D, Teutsch G (1994) RU 58 668, a new pure antiestrogen inducing a regression of human mammary carcinoma implanted in nude mice. J Steroid Biochem Mol Biol 48:187–196

Vaux DL, Haecker G, Strasser A (1994) An evolutionary perspective of apoptosis. Cell 76:777–779

Vignon F, Rochefort H (1987) Autocrine regulation of breast cancer cell growth by estrogen-induced secreted proteins and peptides. In: Moudgil VK (ed) Recent advances in steroid hormone action. De Gruyter, Berlin, pp 405–425

Vignon F, Bardon S, Chalbos D, Rochefort H (1982) Antiestrogenic effect of R5020, a synthetic progestin in human breast cancer cells in culture. J Clin Endocrinol Metab 56:317–323

Vignon F, Capony F, Chambon M, Freiss G, Garcia M, Rochefort H (1986) Autocrine growth stimulation of the MCF7 breast cancer cells by the estrogen-regulated 52 K protein. Endocrinol 118:1537–1545

Vignon F, Bouton MM, Rochefort H (1987) Antiestrogens inhibit the mitogenic effect of growth factors on breast cancer cells in the total absence of estrogens. Biochem Biophys Res Commun 146:1502–1508

Wakeling AE, Bowler J (1987) Steroidal pure antiestrogens. J Endocrinol 112:R7-R10

Wakeling AE, Bowler J (1992) ICI 182,780, a new antiestrogen with clinical potential. J Steroid Biochem Mol Biol 43:173–177

Wakeling AE, Newboult E, Peters SW (1989) Effects of antiestrogens on the proliferation of MCF7 human breast cancer cells. J Mol Endocrinol 2:225–234

Wärri AM, Huovinen RL, Laine AM, Martikainen PM, Härkönen PL (1993) Apoptosis in toremifene-induced growth inhibition of human breast cancer cells in vivo and in vitro. J Natl Cancer Inst 85:1412–1418

Westley B, Rochefort H (1980) A secreted glycoprotein induced by estrogen in human breast cancer cell lines. Cell 20:352–362

Wijsman JH, Jonker RR, Keijzer R, Van de Velde CJH, Cornelisse CJ, Van Dierendonck JH (1993) A new method to detect apoptosis in paraffin sections: in situ end-labeling of fragmented DNA. J Histochem Cytochem 41:7–12

Wyllie AH (1980) Glucocorticoid-induced thymocyte apoptosis is associated with endogenous nuclease activations. Nature 284:555–556

9 Differentiation and Apoptosis as a Therapeutic Strategy for Hormone-Dependent Cancers

H. Michna and K. Parczyk

Our research efforts have been aimed at identifying a compound for the treatment of hormone-dependent cancers. For this purpose the progesterone antagonist onapristone has been selected (Neef et al. 1983; Michna et al. 1989a; Schneider et al. 1989) and is now in a phase III clinical trial. The ability of this compound class to prevent the growth of mammary carcinomas requires the presence of progesterone receptors within the tumors (Michna et al. 1989b). These compounds neither block the secretion of pituitary or ovarian hormones nor are they cytotoxic drugs in experimental rodent models. In addition, the tumor inhibitory potential of progesterone antagonists has proven to be independent of their antihormonal (= antiprogestational) activity and these compounds exert their tumor-inhibiting potential even in the absence of progesterone (Michna et al. 1989b).

Considering the physiological function of the progesterone receptor, which is the mediation of differentiation, it seemed expedient to analyze the morphology of mammary carcinomas after treatment with progesterone antagonists. From comparative studies in MXT, DMBA, and NMU breast cancer models we draw the following conclusions:

In experimental breast cancer models we detected that the tumor inhibition of progesterone antagonists, such as onapristone was accompanied by a shift of undifferentiated tumor epithelial cells towards glandular structures.

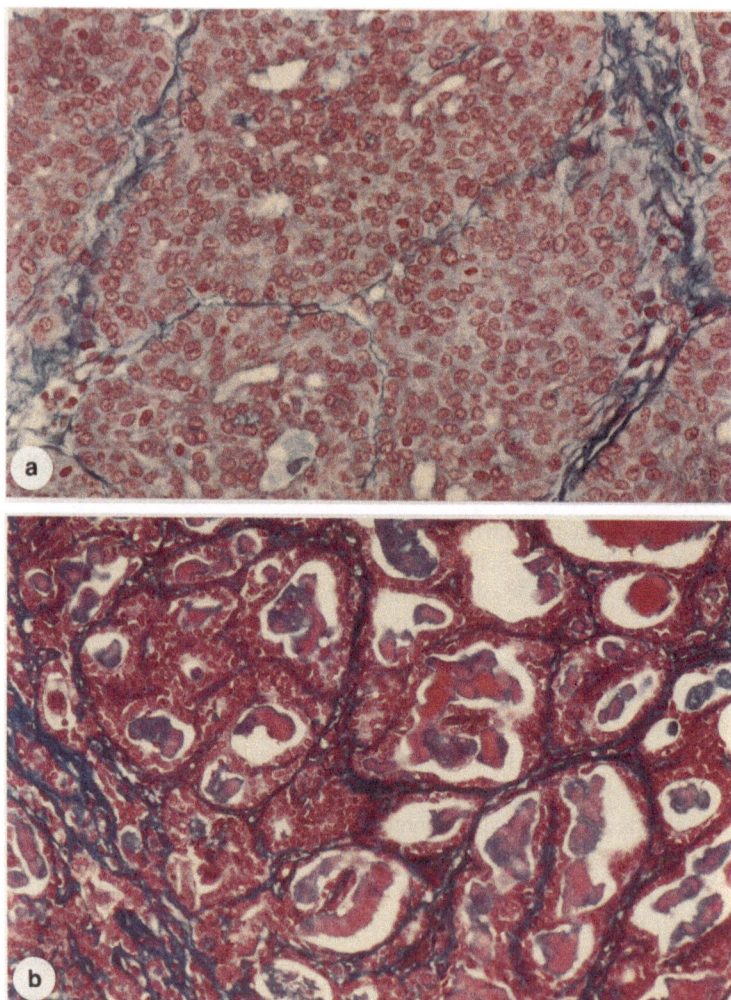

Fig. 1a,b. DMBA-induced mammary carcinoma of the rat; light microscopic characteristics of untreated control and progesterone antagonist (onapristone)-treated tumors after azane staining. It is obvious that after treatment with onapristone more dysplastic glandular structures are filled with secretory material, ×180

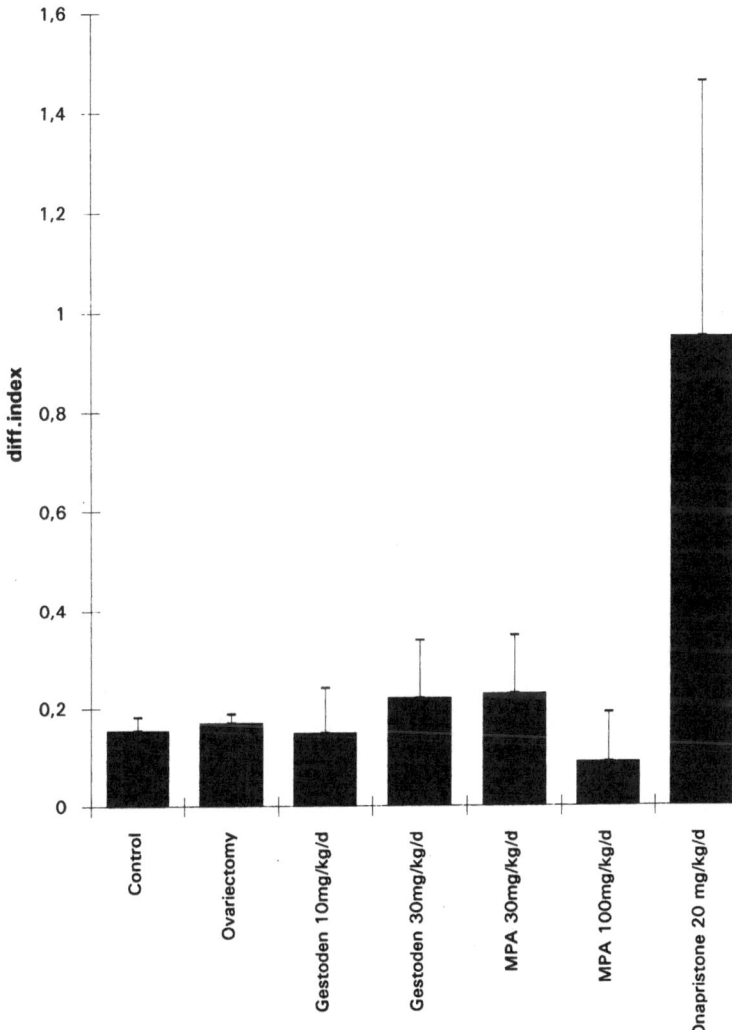

Fig. 2. Using morphometrical procedures the volume density of glandular structures and undifferentiated, spindle-shaped tumor epithelial cells was estimated in DMBA-induced mammary carcinomas and used to calculate a "differentiation index"; no effect was seen after treatment with high doses of progestins, such as gestoden and medroxyprogesteroneacetate (*MPA*), or after ovariectomy, whereas treatment with an antiprogestin significantly enhanced the differentiation index

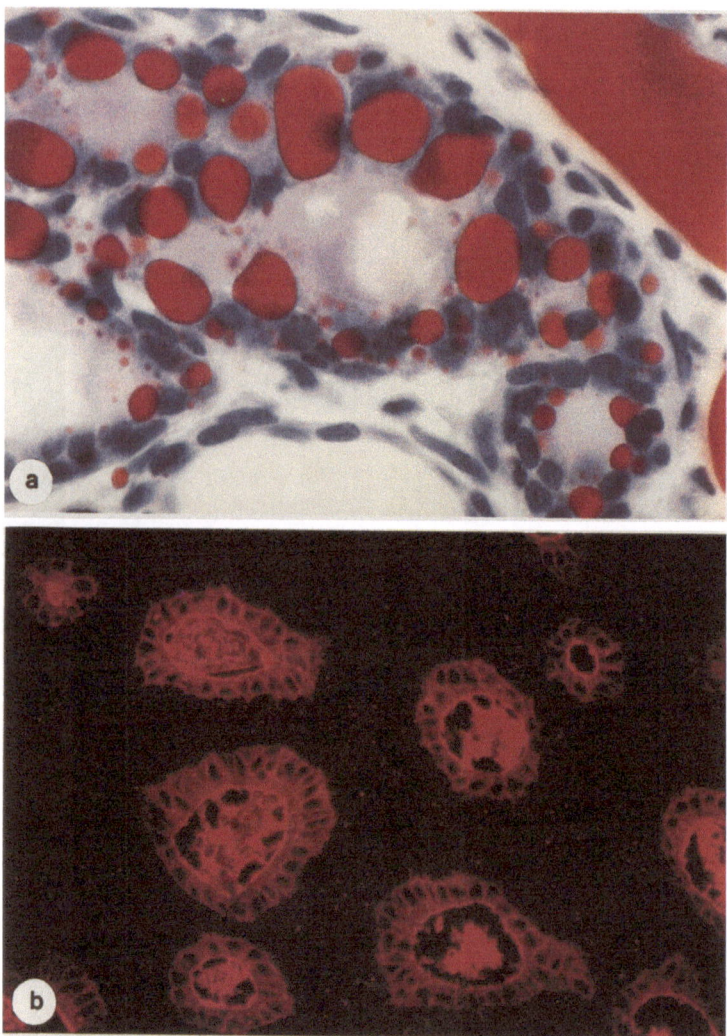

Fig. 3. a Staining of progesterone antagonist (onapristone) treated-DMBA induced mammary carcinomas with "oil-red" revealed the secretion of fatty
acids; the secreted material is stained in red, ×920. **b** In addition, fluorescence
immunohistochemistry detected the secretion of casein, ×320

This reaction pattern was already obvious at the qualitative level of light microscopy (Fig. 1). We could also ensure this reaction by using morphometrical procedures introducing a new differentiation factor, estimating and bringing into relation the amount of undifferentiated, spindle-shaped cells to differentiated tumor epithelial cells arranged to dysplastic, ductlike structures (Gehring et al. 1991). A higher degree of differentiation of hormone-dependent mammary carcinomas (MXT, NMU, and DMBA tumors) was only detected after treatment with progesterone antagonists, whereas ovariectomy, treatment with tamoxifen, or high doses of estrogens or progestins did not influence the degree of differentiation (Fig. 2). The same conclusion could be drawn using classical grading systems such as the WHO or Richardson grading system (Gehring et al. 1991). In line with the above, our data indicate that only after treatment with progesterone antagonists did the tumor epithelial cells show clear evidence of secretory activation – mammary carcinoma cells secrete fatty acids (Fig. 3a) and casein (Fig. 3b). In addition to secretion, transmission electron microscopy revealed apoptotic tumor epithelial cells in breast tissue (Kerr et al. 1972; Schweichel and Merker 1973; Bursch et al. 1985), whereas after ovariectomy necrobiotic tumor cells dominate the picture (Fig. 4).

The estimation of indices of mitotic and apoptotic tumor cell reactions revealed that all endocrine treatment strategies of breast cancer, such as ovariectomy, treatment with tamoxifen, treatment with the pure antiestrogen ICI 164 384 (Wakeling and Bowler 1987), or treatment with high doses of estrogen or progestins may enhance the degree of apoptosis; the most significant apoptotic reactions were observed after treatment with an antiprogestin (Fig. 5), whereas no significant differences were noted in mitogenic reactions. These data are in agreement with recent reports indicating that tamoxifen derivatives may induce apoptotic cell death in mammary cancer models (see overview in Vignon and Rochefort 1994).

Since it was also shown earlier that tamoxifen may stimulate the secretion of the "negative" growth factor transforming growth factor-β_1 (TGF-β_1) in breast cancer tissue of patients treated with tamoxifen (Colletta et al. 1992) and that (liver) cells undergoing active cell death by apoptosis secrete TGF-β_1 (Bursch et al. 1993), we studied the expression of TGF-$\beta_{1,2,3}$ within the experimental mammary carcinomas at the mRNA level and protein level. It was only at the protein level that

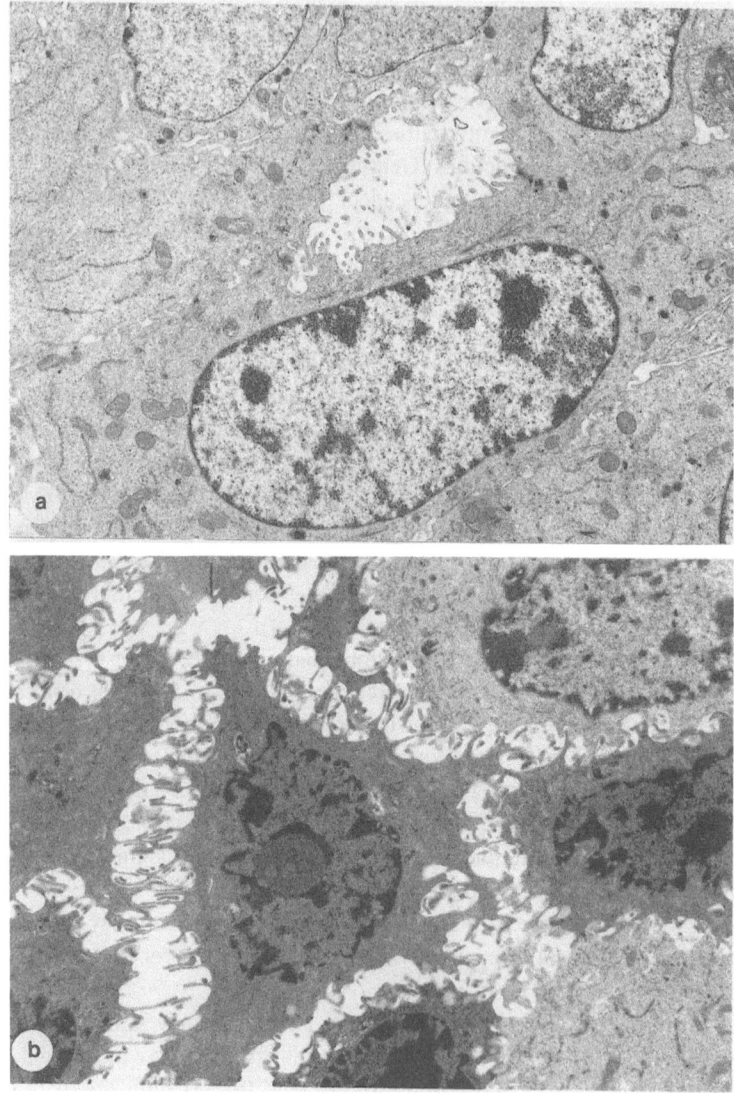

Fig. 4a–d. Transmission electron microscopy of MXT mammary carcinomas of the mouse. Whereas after ovariectomy the tumors display necrobiotic reactions (**b**), in comparison to control tumors grown on intact animals (**a**) after treatment with an antiprogestins, the tumor cells are arranged to glandular

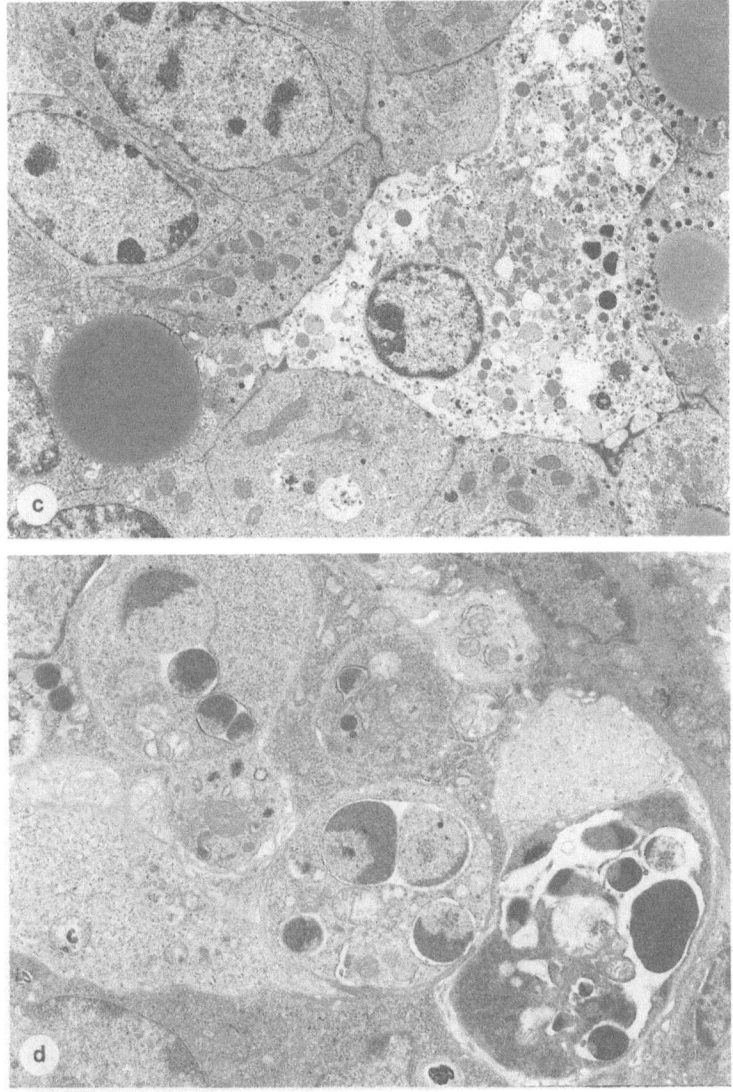

structures bearing signs of secretory activity (**c**). They contain cells undergoing apoptotic cell death (**d**) expressing the characteristic apoptotic nuclear, heterochromatine dense material, ×9600

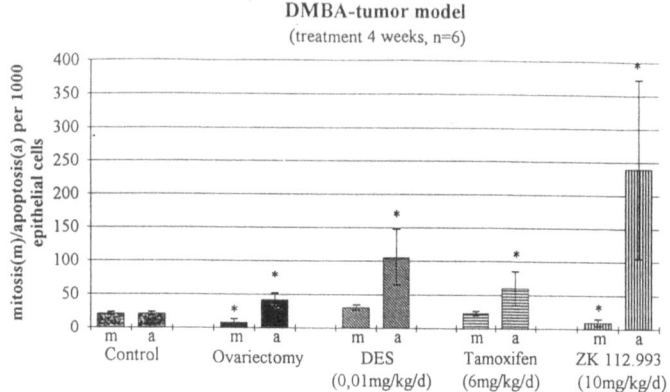

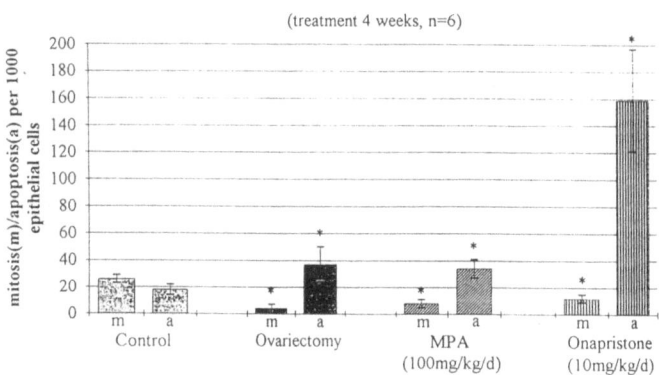

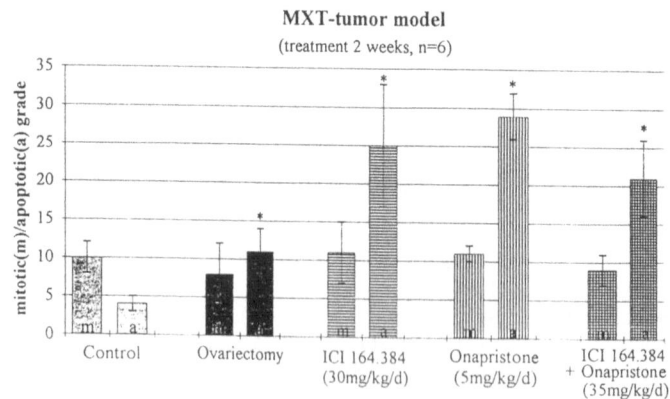

Table 1. TGF β1: immunhistochemical scoring

EO 91-0416 MNU tumor model (treatment 4 weeks)	Stromal tissue	Epithelial tissue	Glandular lumina
Control ($n = 9$; $N = 13$)	3.0 (±0.6)	2.2 (±0.9)	2.0 (±0.6)
Ovariectomy ($n = 4$; $N = 4$)	1.4 (±0.5)	1.8 (±0.6)	2.5 (±0.6)
Tamoxifen ($n = 6$; $N = 10$) (5 mg/kg per day)	3.7* (±0.7)	2.6 (±0.6)	2.1 (±0.6)
Onapristone ($n = 13$; $N = 14$) (20 mg/kg per day)	1.6* ** (±0.8)	3.5* ** (±0.8)	2.9* ** (±0.8)

* Significant difference to control ($p < 0.05$; KRWA test); ** significant difference to Tamoxifen ($p < 0.05$; KRWA test).

we could detect a higher degree of TGF-β_1 secretion by immunohistochemical examination after treatment with progesterone antagonists (Fig. 6a,b), and the degree of staining with the proliferation marker PCNA was reduced simultaneously (Fig. 6a,b). Whereas after treatment with tamoxifen a higher degree of TGF-β_1 immunostaining was localized in stromal cells, after treatment with progesterone antagonists such as onapristone higher staining was displayed by tumor epithelial cells as assessed by semiquantitative analysis (Table 1). Finally, these data are in agreement with in vitro observations, indicating that TGF-β is a hormonally regulated negative growth factor in human breast cancer cells (Knabbe et al. 1987; Jeng and Jordan 1991).

Fig. 5. Morphometric analysis of the mitotic and apoptotic reaction in DMBA-induced and MXT mammary carcinomas after blockade of tumor growth with different hormone therapies; most significant apoptotic reactions after treatment with the progesterone antagonists onapristone or ZK 112.993; nevertheless the pure antiestrogen ICI 164.384 and also the other less pronounced hormone regimen may influence apoptosis when given in high doses. *DES*, diethylstilbestrol; *MPA*, medroxyprogesteronacetate

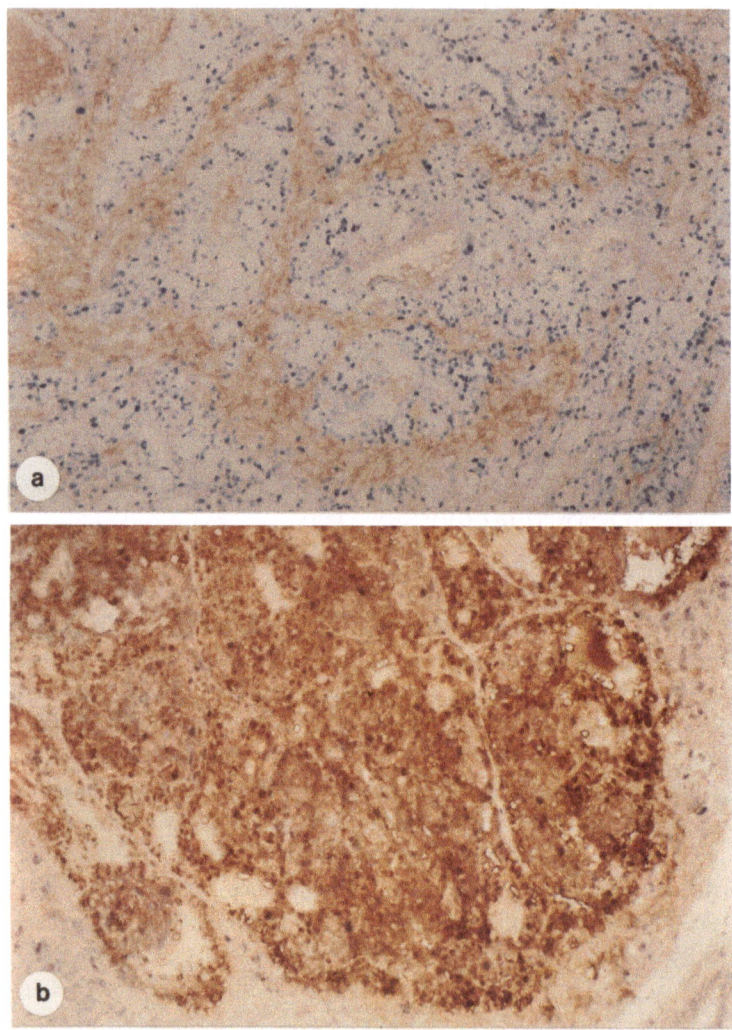

Fig. 6a,b. Coexpression studies by use of immunohistochemistry of the growth factor TGF-β and the proliferation marker proliferating cell nuclear antigen (PCNA); reduction of PCNA staining and enhanced epithelial localization of TGF-β after treatment of DMBA-induced mammary carcinomas with the antiprogestin onapristone, ×90

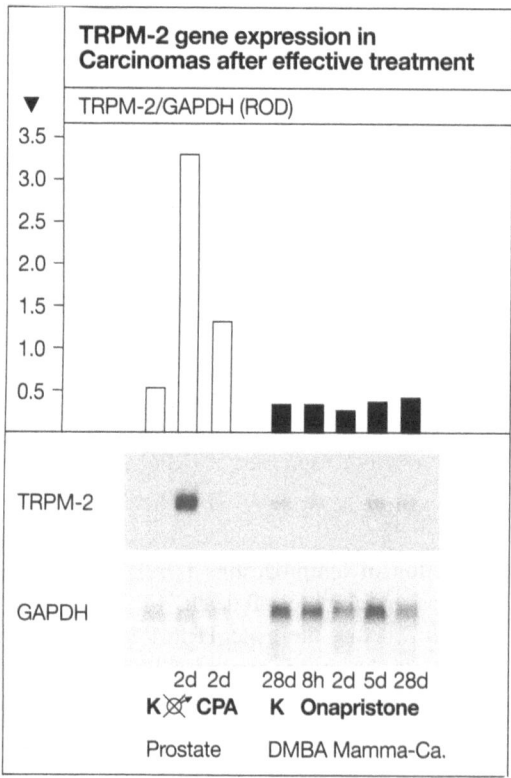

Fig. 7. Northern blot analysis of TRPM-2 gene expression in rat tissue. Effects of orchiectomy or treatment with the antiandrogen cyproterone acetate (*CPA*; 10 mg/kg per day) on the rat prostate. Effects of onapristone (5 mg/kg per day) on DMBA-induced mammary carcinomas after 8 h, 2, 5, and 28 days of treatment. The samples were pooled from five individual prostates/tumors

Interestingly, observations in some stem cell types indicate that hormonal control of cell differentiation and and cell cycle changes are somehow linked (Walker et al. 1993), and differentiation-specific arrest has already been proposed as this link. Keeping this in mind, the accumulation of the tumor cells in G_0/G_1 of the cell cycle after treatment of hormone-dependent breast cancers in vivo with progesterone antagonists (Michna et al. 1990) lends further support to the concept that

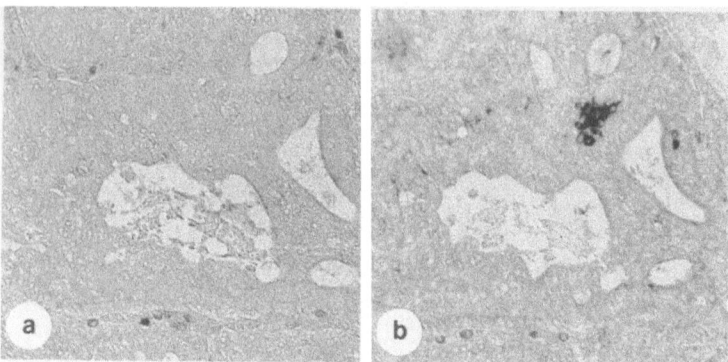

Fig. 8a,b. Immunhistochemical staining of TRPM-2 in DMBA-induced mammary carcinomas after treatment with onapristone. **a** IgG control without the TRPM-2 antibody; note focal expression of TRPM-2, ×180

the antitumor action of antiprogestins is related to the induction of differentiation leading to active cell death.

It has been shown that the gene TRPM-2 (testosterone repressed prostate message) or clusterin is expressed in a wide variety of tissues undergoing active cell death (Montpetit et al. 1986; Rennie et al. 1988; Tenniswood et al. 1990), especially secretory epithelial cells in the prostate and mammary gland, although the functional significance of TRPM-2 is still under investigation (Wilson et al. 1994). Unlike an enhanced expression of the TRPM-2/clusterin gene, after orchiectomy in regressing normal prostate tissue (Fig. 7) as was originally discovered by Léger et al. (1988), we could not detect an upregulation either by ovariectomy or treatment with progesterone antagonists in the analysis of the mRNA levels in homogenates of regressing DMBA-induced mammary carcinomas (Fig. 7). Nevertheless, the immunohistochemical findings of the expression of the clusterin/TRPM-2 gene have revealed that the focal expression is enhanced in regressing mammary carcinomas after treatment with antiprogestins (Fig. 8).

Apoptotic cells similar to those seen in mammary carcinoma epithelial cells were also detected in progesterone receptor-positive uterine epithelial cells (Michna et al. 1989b). Most interestingly, there was also an increase in apoptotic cell death in the absence of progesterone in

Table 2. Morphological reactions in ovarrectomized, mamonary carcinoma-bearing mice treated with estradiol benzoate and simultaneously treated with different progesterone antagonists

Treatment	Incidence of apoptotic nuclei [%]	Morphology of epithelium
OV + EB[a]	15 ± 5 (n = 4)	Single layer
OV + EB + ZK 98.299 (10 mg/kg per day)	42 ± 6 (n = 4)	Multirowed, multilayered
OV + EB + RU 486 (10 mg/kg per day)	32 ± 9 (n = 6)	Multirowed, multilayered

Note enhanced appearance of epithelial cells undergoing apoptotic cell death and differentation of the superficial uterine epithelium from single-layered to multirowed and even multilayered epithelium.
EB, estradiol benzoate; OV, ovarectomized mice.
[a]Application of compounds all subcutaneously.

ovariectomized estradiol-treated animals (Table 2), which is also associated with the expression of the clusterin/TRPM-2 gene (Fig. 9). Thus, the progesterone antagonists specifically interact with the active cell death pathway of epithelial cells without killing stromal cells in tumors or uteri. Nevertheless, our studies on the marker glycoprotein tenascin have revealed the involvement of stromal reactions during tumor regression by the mechanism apoptosis induced by antiprogestins (Vollmer et al. 1992). These and other data (Cunha et al. 1987) suggest that stromal–epithelial interactions appear to be critical both for the mechanism of active cell death as well as cell survival.

These data have led us to propose that progesterone receptor antagonists, such as onapristone, use the physiological function of the progesterone receptor and mediate differentiation and stimulate genes implicated in active cell death, finally leading to apoptosis. These data also favor the idea that specific growth factors, such as TGF-β, may cause epithelial cell death (Kyprianou and Isaacs 1989; Warner et al. 1991; Chung et al. 1992). However, the fact that progesterone antagonists are

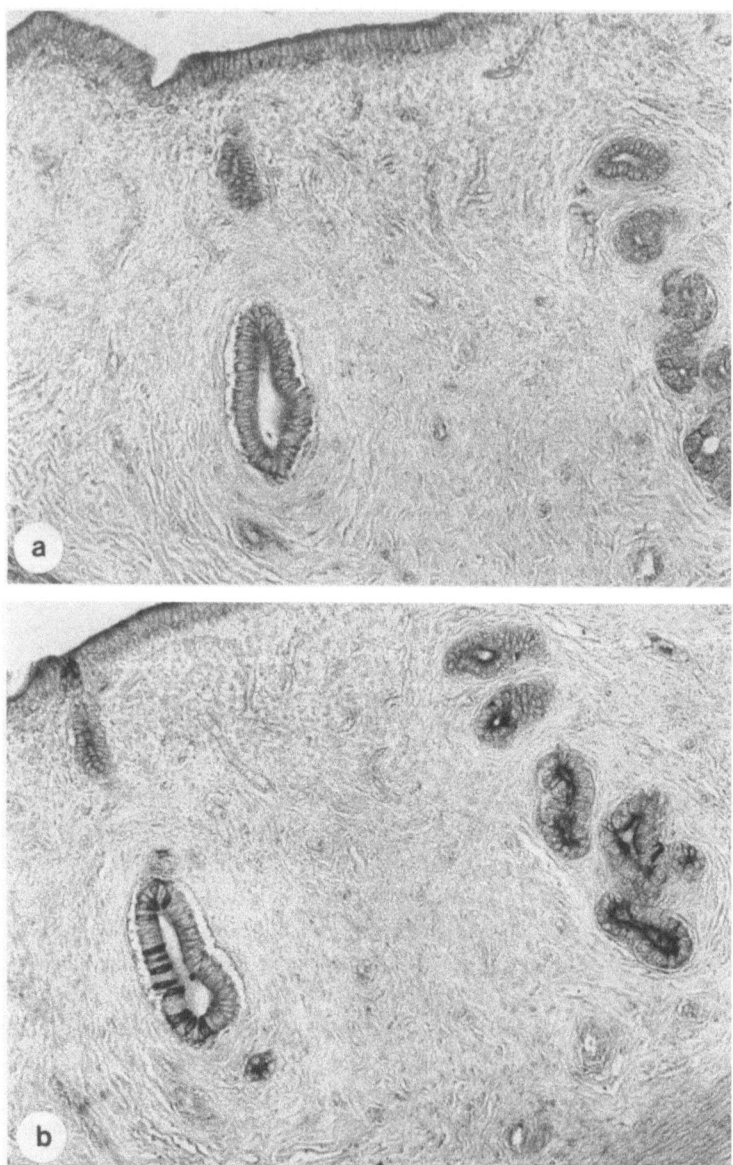

Fig. 9a,b. Legend see p. 175

able to exert these effects even in the absence of progesterone still raises the question as to how the binding of the antiprogestins to the unliganded progesterone receptor can produce the pronounced effects of differentiation and apoptosis which result in strong tumor inhibition.

Most astonishingly, in addition to progesterone antagonists, androgen receptor antagonists may also induce apoptotic cell death in regressing prostates, although this process is judged to be a dedifferentiation of the prostate tissue (Aumüller 1979). It thus appears that many other antihormones, tamoxifen-like antiestrogens (Kyprianou et al. 1991; Vignon and Rochefort 1994), and vitamin D3 analogues (Chap. 10, this volume), as well as chemotherapeutic agents (Walker et al.1994) are capable of inducing apoptosis independent of their effect on the degree of differentiation. However, the progesterone antagonists are the only compound class which, in our hands, seem to be capable of triggering apoptosis and differentiation in experimental mammary carcinomas in vivo. It would be interesting to know whether both of these mechanisms play a role in the clinical use of antiprogestins in breast cancer treatment.

In terms of using drugs which induce apoptotic cell death as a therapeutic principal one needs to consider the issue of possible side effects and the goal to achieve selectivity in killing tumor cells. There are data indicating that, possibly based on high levels of oncogene expression of tumor cells (Vaux et al. 1988; Reed et al. 1990), malignant cells seem to be more likely to be susceptible to undergoing apoptosis than normal cells, which has already been considered to be the price these cells have to pay for their proliferative advantage (Martin and Cotter 1994).

An exciting strategy for the treatment of cancer, especially hormone-independent cancers, would therefore be to create agents that can interact with the cell suicide mechanism of neoplastic cells. The most crucial question then is the choice of a suitable target for tumor-selective cell death induction: Theoretically, either one could consider activating factors *inducing* cell death or the blockade of a factor that *protect* from cell death (e.g., blockade of bcl-2 gene expression or protein function).

◀ **Fig. 9a,b.** Focal expression of TRPM-2 in uterine epithelial cells after treatment with progesterone antagonists (**b**); the IgG control is shown in **a**

Table 3. Proteins known to promote and prevent apoptotoc cell death and which, therefore, may serve as targets for a therapeutic strategy based on apoptosis

Proteins involved in cell death		Function
ICE (CED-3)	Promotes death	Il-Iβ converting enzyme (protease)
TRPM-2	Promotes/prevents death?	Manifold
BCL-2 (CED-9)	*Prevents* death	Opposes/binds BAX; blocks formation of oxygen radicals
BAX	Promotes death	Opposes/binds BCL-2
p53	Promotes death	Following DNA damage; guardian of the genome
FAS/APO1-ligand	Promotes death	Binds to membrane receptor for cell death command; receptor belongs to NGF/TNF familiy
BCL-X long	*Prevents* death	Opposes/binds BCL-X short
BCL-X short	Promotes death	Opposes/binds BCL-X long
c-Myc	Promotes death	Promotes even proliferation
TNF	Promotes death	*Ligand*
TGF-β	Promotes death	*Ligand*
Estrogen/androgen/gestagens	*Prevent* death	Stimulate proliferation of hormone dependent tissues

The approach to induce cell death could either be achieved by the design of genuine cell death genes or by identification of a ligand that activates a receptor for cell death command (e.g., FAS/APO-1 ligand). When reviewing the targets involved in cell death there seem to be not only a growing number of proteins involved in cell death (Table 3), but it also appears that their function may also be dependent on interactions (Bissonette et al. 1994), most probably within a cascade system. Nevertheless, the induction of active cell death by apoptosis may provide a new effective strategy for the treatment of even hormone-relapsed tumors (Kyprianou et al. 1994).

As for the clinical use of progesterone antagonists, one can conclude based on the data presented here that, a century after Schinzinger reported on the beneficial outcome of ovarian hormone depletion after ovariectomy (Schinzinger 1889), an innovative hormone treatment represents a promising future.

Acknowledgements. The following colleagues provided us with materials, for which we are very grateful: L. Wakefield with antibodies against TGF-$\beta_{1,2,3}$; M. Tenniswood with antibodies against TRPM-2; M. Grosse with antibodies against casein. We also cordially thank S. Gehring, Y. Nishino, M.R Schneider, and G. Vollmer, who contributed to some of the work discussed in this article.

References

Aumüller G (1979) Prostate gland and seminal vesicles. In: Oksche A, Vollrath L (eds) Handbuch der mikroskopischen Anatomie des Menschen, vol VII/6. Springer, Berlin Heidelberg New York

Bissonette, Shi Y, Mahboubi A, Glynn JM, Green DR (1994) c-myc and apoptosis. In:Tomei LD, Cope FO (eds) Apoptosis II: the molecular basis of apoptosis in disease. Cold Spring Harbor Laboratory Press, Cold Spring Harbor

Bursch W, Taper HS, Laur B, Schulte-Hermann R (1985) Quantitative histological and histochemical studies on the occurrence and stages of controlled cell death (apoptosis) during regression of the rat liver hyperplasia. Virchows Arch [B] 50:153–166

Bursch W, Oberhammer F, Jirtle RL, Askari M, Sedivy R, Grasl-Kraupp B, Purchio AF (1993) Transforming growth factor-$_1$ as a signal for induction of cell death by apoptosis. Br J Cancer 67:531–536

Chung LW, Li W, Gleave ME, Hsieh JT, Wu HC, Sikes RA, Zhau HE, Bandyk MG, Logothetis CJ et al (1992) Human prostate cancer model: roles of growth factors and extracellullar matrices. J Cell Biochem Supp 16H

Colletta AA, Wakefield LM, Howell FV, van Roozendaal KE, Danielpour D, Ebbs SR, Sporn MB, Baum M (1992) Anti-estrogens induce the secretion of active transforming growth factor-beta from fetal fibroblasts. Br J Cancer 62:405–409

Cunha GR, Donjacour A, Cooke PS, Mee S, Bigsby RM, Higgins SJ, Sugimura Y (1987) The endocrinology and developmental biology of the prostate. Endocr Rev 8:338–362

Gehring S, Michna H, Kühnel W, Nishino Y, Schneider MR (1991) Morphometrical and histochemical studies on the differentiation potential of progesterone antagonists in experimental mammary carcinomas. Acta Endocrinol (Copenh) 124:89

Jeng MH, Jordan VC (1991) Growth stimulation and differential regulation of transforming growth factor-beta$_{1,2,3}$ messenger RNA levels by norethidrone in MCF-7 human breast cancer cells. Mol Endocrinol 5:1120–1128

Kerr JFR, Wyllie AH, Currie AR (1972) Apoptosis: a basic phenomenon with wide ranging implications in tissue kinetics. Br J Cancer 26: 239–257

Knabbe C, Lippman ME, Wakefield LM, Flanders KC, Kasid A, Derynyck R, Dickson RB (1987) Evidence that transforming growth factor-beta is a hormonally regulated negative growth factor in human breast cancer cells. Cell 48:417–428

Kyprianou N, Isaacs JT (1989) Expression of transforming growth factor-β in the ventral prostate during castration-induced programmed cell death. Mol Endocrinol 3:1515–1522

Kyprianou N, English HF, Davidson NE, Isaacs JT (1991) Programmed cell death during regression of the MCF-7 human breast cancer following estrogen ablation. Cancer Res 51:162–166

Kyprianou N, Bains AK, Jacobs SC (1994) Induction of apoptosis in androgen-independent human prostate cancer cells undergoing thymineless death. Prostate 25:12–22

Léger JG, Le Guellec R, Tenniswood MP (1988) Treatment with antiandrogens induces an androgen-repressed gene in the rat ventral prostate. Prostate 13: 131–142

Martin SJ, Cotter TG (1994) Apoptosis of human leukemia: induction, morphology, and molecular mechanisms. In: Tomei LD, Cope FO (eds) Apoptosis II: the molecular basis of apoptosis in disease. Cold Spring Harbor Laboratory Press, Cold Spring Harbor

Michna H, Schneider MR, Nishino Y, El Etreby MF (1989a) Antitumor activity of the antiprogesterones ZK 98.299 and RU 38.486 in hormone dependent rat and mouse mammary tumors: mechanistic studies. Breast Cancer Res Treat 14:275–288

Michna H, Schneider MR, Nishino Y, El Etreby MF (1989b) The antitumor mechanism of progesterone antagonists is a receptor mediated antiproliferative effect by induction of terminal cell death. J Steroid Biochem 34:447–453

Michna H, Schneider MR, Nishino Y, El Etreby MF, McGuire WL (1990) Progesterone antagonists block the growth of experimental mammary tumors in G_0G_1. Breast Cancer Res Treat 17:155–156

Montpetit Ml, Lawless KR, Tenniswood M (1986) Androgen-repressed messages in the rat ventral prostate. Prostate 8:25–36

Neef G, Beier S, Elger W, Henderson, Wiechert G (1983) New steroids with antiprogestational and antiglucocorticoid activities. Steroids 44:349–372

Reed JC, Cuddy M, Haldar S, Croce, Nowell P, Makover D, Bradley K (1990) BCL-2-mediated tumorigenicity of a human T-lymphoid cell line: Synergy with MYC and inhibition by BCL2 antisense. Proc Natl Acad Sci USA 87:3660

Rennie PS, Bruchowsky N, Buttyan R, Benson M, Cheng H (1988) Gene expression during the early phases of regression of the androgen-dependent Shionogi mouse mammary carcinoma. Cancer Res 48:6309–6312

Schinzinger K (1889) Über Carcinoma mammae. Hirschwaldt, Berlin (Verhandlungen der Deutschen Gesellschaft für Chirurgie)

Schneider MR, Michna H, Nishino Y, El Etreby MF (1989) Antitumor activity of the progesterone antagonists ZK 98.299 and RU 38.486 in the hormone-dependent MXT mammary carcinoma model of the mouse and the DMBA- and MNU-induced mammary tumor model of the rat. Eur J Cancer Clin Oncol 25: 691–701

Schweichel JU, Merker HJ (1973) The morphology of various types of cell death in prenatal tissues. Teratology 7:253–266

Tenniswood MP, Montpetit ML, Léger JG, Wong P, Pineaut JM, Rouleau M (1990) Epithelial-stromal interactions and cell death in the prostate. In: Farnsworth WE, Ablin RJ (eds) The prostate as an endocrine gland. CRC Press, Boca Raton, pp 187–207

Vaux DL, Cory S, Adams JM (1988) bcl-2 gene promotes haemopoietic cell survival and cooperates with c-myc to immortalize pre-B cells. Nature 335:440–442

Vignon F, Rochefort H (1994) Anti-growth factor activity of antiestrogens in human breast cancer cells. In: Tenniswood M, Michna H (eds) Apoptosis in hormone dependent cancers. Springer, Berlin Heidelberg New York

Vollmer G, Michna H, Ebert K, Knuppen R (1992) Down-regulation of tenascin expression by antiprogestins during terminal differentiation of rat mammary tumors. Cancer Res 52:4642–4648

Wakeling AE, Bowler J (1987) Steroidal pure antiestrogens. J Endocrinol 112:R7-R10

Walker PR, Kwast-Welfeld J, Gourdeau H, Leblanc J, Neugebauer W, Sikorska M (1993) Relationship between apoptosis and the cell cycle in lymphocytes: role of protein kinase C, tyrosine phosphorylation and AP 1. Exp Cell Res 207:142–151

Walker PR, Testolin L, Armato U, Marceau N, Gourdeau H, Sikorska M (1994) Modulation of oncogenes by apoptosis. In: Tenniswood M, Michna H (eds) Apoptosis in hormone dependent cancers. Springer, Berlin Heidelberg New York

Warner NI, Nelson D, Ludlow J, Scott DW (1991) Antiimmunoglobulin treatment of murine B-cell lymphomas induces active transforming growth factor beta but pRB hypophosphorylation is transforming growth factor beta independent. Cell Growth Differ 3:175179

Wilson MR, Easterbrook-Smith SB, Taillefer D, Lakins J, Tenniswood M (1994) Mechanism of induction and function of clusterin at sites of cell death. In: Harmony J (ed) Clusterin: function in vertebrate organ development, function and adaptation. Landes, Austin (in press)

10 1,25 Dihydroxyvitamin D₃: Coordinate Regulator of Active Cell Death and Proliferation in MCF-7 Breast Cancer Cells

M. Simboli-Campbell and J. Welsh

10.1 Introduction

Maintenance of normal tissue cell number requires a net balance between the rate of cell division and the rate of cell death. The uncontrolled growth characteristic of both benign and metastatic tumors may be due to either an increase in the rate of proliferation or a decrease in the rate of cell death, or both (Williams 1991). Cell death can be classified into one of two categories: necrosis, the result of tissue insult or injury; and apoptosis, or active cell death (ACD), a process of active cellular self-destruction. ACD is an asynchronous process consisting of a series of distinct steps which are common to epithelial cells of the prostate, mammary gland, liver, and many other tissues (Bursch et al. 1990). A schematic representation of the morphology and biochemistry of ACD is shown in Fig. 1. The first visible stage of ACD in regressing

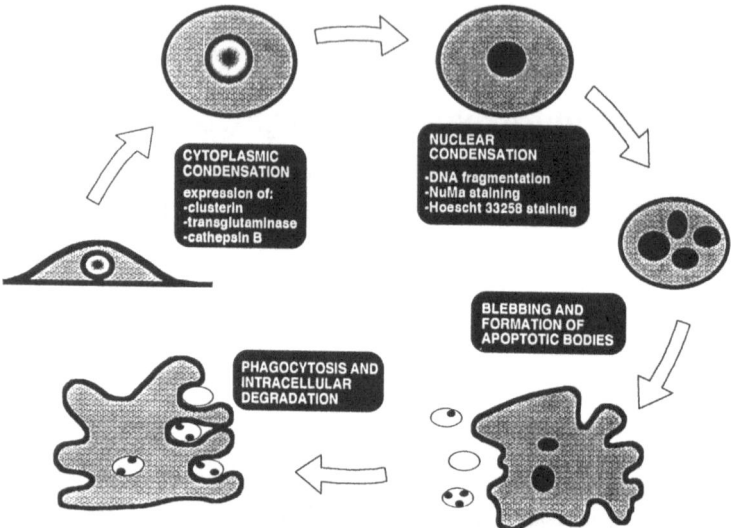

Fig. 1. Morphological and biochemical markers of active cell death. Schematic diagram illustrating characteristic features of active cell death (apoptosis) and biochemical/molecular markers of the process. See text for explanation

mammary gland involves disruption of the cytoskeleton and condensation of intermediate filaments around the nucleus. During this stage, loss of cell to cell contacts results in the release of dying cells from the basement membrane in vivo or rounding up and detachment of adherent cells in vitro. The early apoptotic cell thus lies above the plane of the monolayer and is characterized by cytoplasmic condensation. Chromatin condensation occurs simultaneously, producing the hyperchromatic, pyknotic nucleus, which then separates into discrete masses of condensed chromatin. In some, but not all, forms of ACD, cleavage of DNA into large (50–300-kbp) fragments and further degradation into 200-bp fragments, characteristic of a nucleosome ladder, can be demonstrated (Oberhammer et al. 1993). Finally, the cell itself fragments into a number of membrane-bound vesicles called apoptotic bodies, some of which contain chromatin. In vitro, apoptotic cells and bodies can be recovered in the media, whereas in vivo, these are removed by phagocytic macrophages.

Table 1. Distribution of the vitamin D receptor in human breast and prostate
cancer cell lines

Mammary carcinoma
MCF-7
T47D
ZR-71–1
T-39
BT-20
Prostate carcinoma
Primary cultures
LNCaP
DU-145
PC-3
ALVA-31
TSU-PR1
PPC-1
JCA-1

ACD is associated with increased expression and/or activation of
specific genes and proteins which are required to mediate these events.
For example, during mammary gland regression in vivo, the steady state
levels of certain mRNAs and proteins, including clusterin/TRPM-2,
transforming growth factor-β (TGF-β), transglutaminase, and cathepsin
B, dramatically increase in parallel with ACD (Tenniswood et al. 1992,
1994; Guenette et al. 1994). Clusterin is a widely used marker of ACD
since its expression has been correlated with ACD in a variety of model
systems, including MCF-7 breast cancer cells after estrogen deprivation
or treatment with antiestrogens (Kyprianou et al. 1991; Warri et al.
1993).

The ability of mammary tumors and breast cancer cell lines to
undergo ACD underscores the potential for targeting ACD in the context of cancer therapy. Since many breast tumors are estrogen receptor
positive, hormone ablation therapies (antiestrogens, antiprogestins) are
currently being used in breast cancer patients to induce ACD in hormone-dependent cells. Although these therapies usually result in significant tumor regression in the short term, the high rate of tumor recurrence is thought to reflect the survival of hormone-independent cells

which do not undergo ACD following hormone ablation and thus continue to proliferate. Since these hormone-independent cells retain the capacity to undergo ACD in response to other inducers such as tumor necrosis factor (TNF) and TGF-β, it is important to identify agents which activate the cell death pathway via mechanisms distinct from that of antiestrogens.

1,25 Dihydroxyvitamin D_3 [1,25(OH)$_2D_3$], the biologically active form of vitamin D_3 (cholecalciferol), is a potent negative growth regulator of breast cancer cells both in vivo and in vitro (Colston et al. 1992). The vitamin D receptor, a member of the steroid/thyroid/retinoic acid family of nuclear receptors, is present in normal and lactating mammary gland as well as in both human and animal mammary tumors (Berger et al. 1987; Colston et al. 1988; Sahota et al. 1991). Table 1 contains a partial list of established cell lines derived from mammary carcinomas which are vitamin D receptor positive. Also listed are prostate cancer cell lines which express the vitamin D receptor, since recent evidence indicates that 1,25(OH)$_2D_3$ can inhibit growth of prostate as well as breast carcinoma cells.

Although it is known that 1,25(OH)$_2D_3$ inhibits growth of both estrogen receptor-positive and estrogen receptor-negative breast cancer cells (Chouvet et al. 1986; Abe et al. 1992), the precise mechanism of its effects remains unclear. To complement data which demonstrate antiproliferative effects of 1,25(OH)$_2D_3$ in breast cancer cells, we have assessed indices of ACD in MCF-7 cells treated with 1,25(OH)$_2D_3$ for up to 96 h. In the studies described here, we demonstrate that inhibition of MCF-7 breast cancer cell growth by 1,25(OH)$_2D_3$ is associated with morphological and biochemical changes indicative of ACD. Our data indicate that ACD is also induced in MCF-7 cells treated with EB 1089, a novel nonhypercalcemic analog of 1,25(OH)$_2D_3$ under development by LEO pharmaceuticals (Ballerup, Denmark) as a chemotherapeutic agent for human breast cancer. These studies provide a firm rationale for investigating the efficacy of novel vitamin D analogs in induction of the cell death pathway in breast tumors.

10.2 Results and Discussion

10.2.1 Evidence for Induction of ACD by 1,25(OH)$_2$D$_3$

10.2.1.1 Cell Growth and Cell Cycle Data

MCF-7 cells were grown to subconfluence (approximately 4 days) and fluid-changed to fresh media containing 100 nM 1,25(OH)$_2$D$_3$ or ethanol vehicle and harvested up to 96 h later. The temporal effect of 1,25(OH)$_2$D$_3$ on MCF-7 cell number indicates that 100 nM 1,25(OH)$_2$D$_3$ significantly reduces the cell number within 48 h (Fig. 2A). Further, while cell number in control cultures increases with time up to 96 h, cell number in 1,25(OH)$_2$D$_3$-treated cultures increases only slightly up to 72 h and then plateaus. To assess the contribution of cell proliferation to the reduction in total cell number, [^{3}H]thymidine incorporation was measured in control and 1,25(OH)$_2$D$_3$-treated cultures for up to 96 h following treatment (Fig. 2B). Data for cell proliferation are expressed as cpm thymidine incorporated per microgram DNA to control for cell number. Under the conditions used in our studies, 1,25(OH)$_2$D$_3$ exerts minor effects on MCF-7 cell proliferation at 48 and 72 h, which are not statistically significant. These data are consistent with studies indicating large decreases in cell number but only moderate decreases in [^{3}H]thymidine incorporation in MCF-7 cells treated with 1,25(OH)$_2$D$_3$ (Abe et al. 1991; Chouvet et al. 1986; Frampton et al. 1983). After 96 h of 1,25(OH)$_2$D$_3$ treatment, [^{3}H]thymidine incorporation is significantly lower in treated cultures, suggesting that prolonged exposure to 1,25(OH)$_2$D$_3$ does inhibit MCF-7 cell proliferation. Taken together, the time course data indicate that the initial limited increase in cell number in 1,25(OH)$_2$D$_3$-treated cultures cannot be attributed solely to inhibition of cell proliferation and suggest an effect of 1,25(OH)$_2$D$_3$ on the cell death pathway.

The distribution of MCF-7 cells in G$_0$/G$_1$, S, and G$_2$M phases of the cell cycle was determined by flow cytometric analysis of both adherent and detached cells stained with propidium iodide. These data indicate the major effect of 1,25(OH)2D3 is to induce accumulation of cells in G$_0$/G$_1$, with a minor effect of the percentage of cells in S phase and no effect on the percentage of cells in G$_2$M (Fig. 2C). The nonhypercalcemic analog, EB1089 (generously supplied by Dr. Lise Binderup, LEO Pharmaceuticals, Ballerup, Denmark), elicits effects identical to those

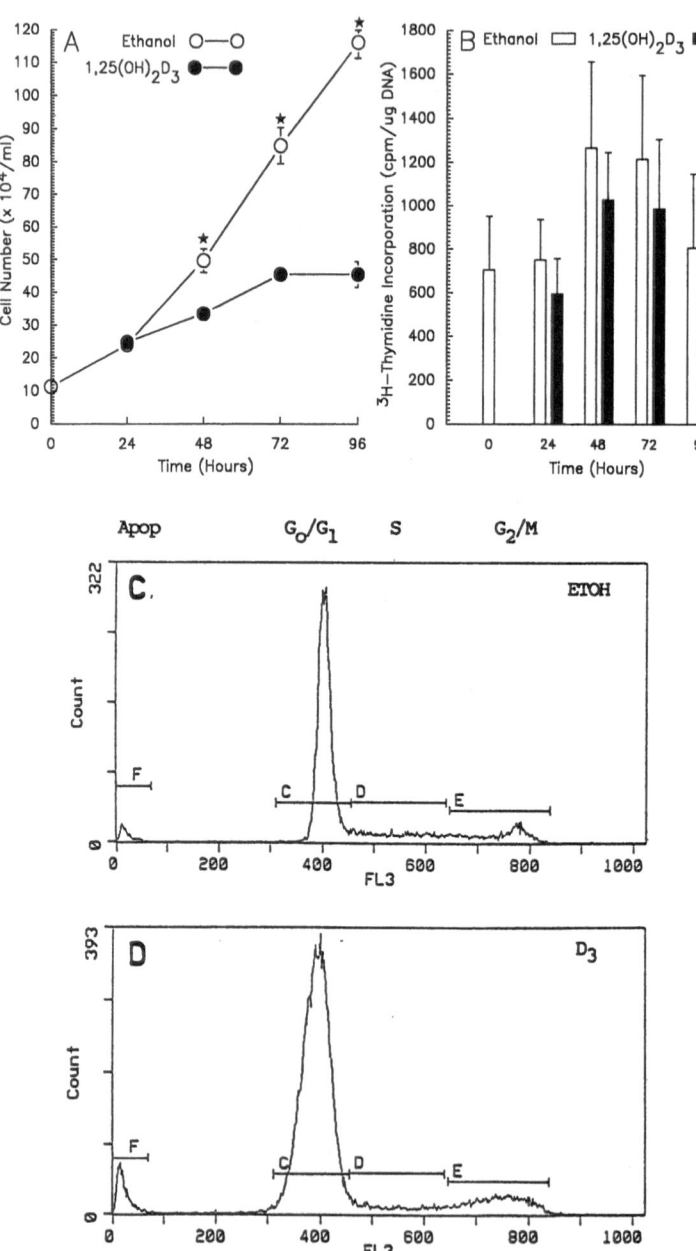

of $1,25(OH)_2D_3$ at 100-fold lower concentrations (M. Simboli-Campbell and J. Welsh, manuscript in preparation). These data are consistent with previous studies on the effects of $1,25(OH)_2D_3$ on cell cycle kinetics of T47D human breast cancer cells (Eisman et al. 1989). Again, it is unlikely that these small effects of $1,25(OH)_2D_3$ on cell proliferation can account for the large differences in cell number between control and $1,25(OH)_2D_3$-treated cultures. Of particular interest is recent evidence which suggests that early G_1 may be the point at which switching between cell cycle progression and induction of ACD occurs (Walker et al. 1993).

10.2.1.2 Morphology, Immunofluorescence, and Biochemistry

We have examined several markers of ACD in order to elucidate the type of cell death induced in MCF-7 cells following $1,25(OH)_2D_3$ treatment. By comparison of phase contrast and Hoescht 33258 staining, apoptotic cells can be easily distinguished from mitotic cells. In cultures treated with ethanol, cells undergoing mitosis are evident under phase contrast, with characteristic mitotic spindles stained by Hoescht 33258 (Fig. 3, top). In contrast, cells treated for 48 h with $1,25(OH)_2D_3$ (Fig. 3, bottom) are shrunken, with condensed cytoplasm, and lie above the adherent cells of the monolayer. These apoptotic cells are readily distinguished from cells undergoing mitosis, as the dying cells display condensed chromatin and irregularly shaped nuclei. Higher magnification (Fig. 4) of an early apoptotic cell in $1,25(OH)_2D_3$-treated cultures clearly shows the loss of cell–cell contacts, the shrunken and condensed cytoplasm, vesicle formation, and blebbing of apoptotic bodies from the dying cell. Parallel Hoescht staining emphasizes the irregularly shaped nucleus and condensed chromatin.

◀ **Fig. 2A–D. A,B** Indices of MCF-7 cell growth following treatment with ethanol (control) or 100 nM $1,25(OH)_2D_3$. MCF-7 cells were treated with 100 nM $1,25(OH)_2D_3$ or ethanol vehicle for up to 96 h prior to determination of cell number (**A**) and [³H]thymidine incorporation (**B**). Results are presented as means ± SEM, n=6. $^*p < 0.05$, control vs. $1,25(OH)_2D_3$ treated. **C** Cell cyle distribution of MCF-7 cells treated for 48 h with ethanol (ETOH, *top*) or $1,25(OH)_2D_3$ (D₃, *bottom*). Cells were fixed, permeabilized, and incubated with propidium iodide prior to analysis of cell cycle phase distribution by flow dytometry. Histograms are plotted as cell counts (*y-axis*) vs. propidium iodide fluorescence (FL3, *x-axis*)

Phase   Contrast   Hoescht   33258

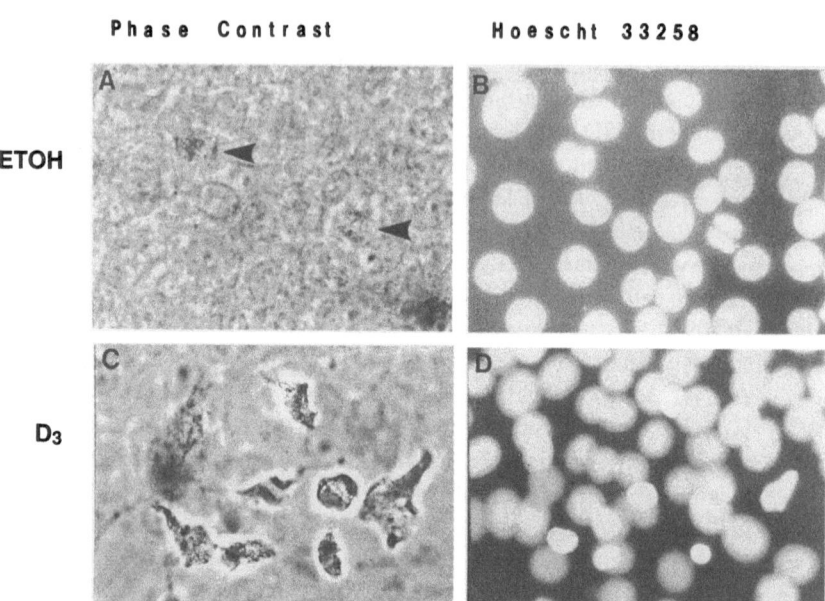

Fig. 3A–D. Morphology and DNA conformation in MCF-7 cells treated with ethanol or 1,25(OH)$_2$D$_3$: low magnification. MCF-7 cells were grown on coverslips, treated with ethanol or 1,25(OH)$_2$D$_3$ for 48 h, fixed and incubated with Hoescht 33258 (0.5 µg/ml) for 1 h. **A,C** phase contrast; **B,D** Hoescht fluorescence. **A,B** ethanol control; **C,D** 1,25(OH)$_2$D$_3$ treated. In **A**, *arrowheads* indicate mitotic cells

Alterations in nuclear architecture during ACD include not only chromatin condensation but reorganization of the nuclear matrix. Condensation of nuclear matrix proteins has been described in MCF-7 cells undergoing ACD in response to TNF or serum deprivation (Miller et al. 1994). Since these changes are distinct from those occurring during mitosis, patterns of nuclear matrix proteins can be used to distinguish mitotic and apoptotic cells. We examined immunofluorescent patterns for nuclear mitotic apparatus protein (NUMA; Yang et al. 1991) to assess the effects of 1,25(OH)$_2$D$_3$ on nuclear matrix organization in MCF-7 cells. NUMA is a nuclear matrix protein associated with the mitotic apparatus. In control cells, diffuse nuclear immunoreactivity is evident in quiescent cells, and in proliferating cells, NUMA reorganizes

Phase Contrast   **Hoescht 33258**

ETOH

D$_3$

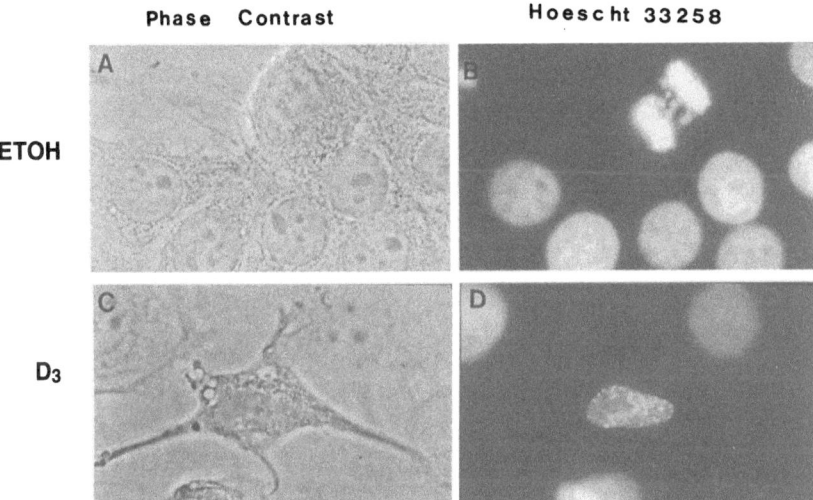

Fig. 4A–D. Morphology and DNA conformation in MCF-7 cells treated with ethanol or 1,25(OH)$_2$D$_3$: high magnification. MCF-7 cells were grown on coverslips, treated with ethanol or 1,25(OH)$_2$D$_3$ for 48 h, fixed and incubated with Hoescht 33258 (0.5 μg/ml) for 1 h. **A,C** phase contrast; **B,D** Hoescht fluorescence. **A,B** ethanol control; **C,D** 1,25(OH)$_2$D$_3$ treated

along the mitotic spindles (Fig. 5A,B). In 1,25(OH)$_2$D$_3$-treated cells exhibiting early signs of apoptosis (Fig. 5c,d), NUMA condensation which is clearly distinct from the pattern in mitotic cells is evident. These changes parallel the changes in chromatin organization observed with Hoescht 33258 (compare to Figs. 3,4) and are similar to the correlation between NUMA immunofluoresence and diamidino-2-phenylindole (DAPI) staining of DNA in MCF-7 cells following adriamycin treatment (Miller et al. 1994).

Cathepsin B is a secreted lysosomal cysteine protease which increases dramatically during ACD in regressing mammary gland (Guenette et al. 1994). To determine whether the intracellular vesicles observed in apoptotic cells represent autophagic lysosomes, we assessed the expression of the cathepsin B protein in control and 1,25(OH)$_2$D$_3$-treated cells by immunofluorescence and western blotting. Both techniques demonstrate a definitive increase in cathepsin B within 48 h of

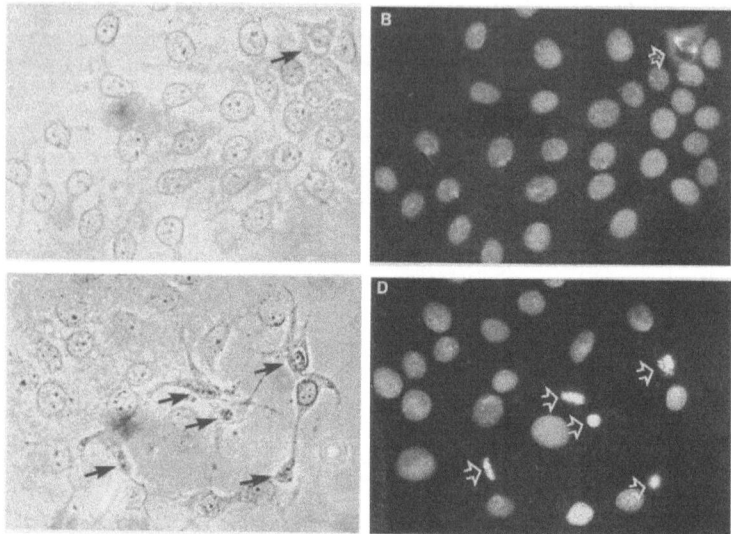

Fig. 5A–D. Nuclear mitotic apparatus protein (NUMA) immunofluorescence in MCF-7 cells treated with ethanol or 1,25(OH)₂D₃. MCF-7 cells grown on coverslips were treated with 100 nM 1,25(OH)₂D₃ or vehicle for 48 h, fixed, and processed for immunofluorescence using a monoclonal antibody to NUMA (MAB Matritech, Inc, Cambridge, MA). **A,C** are phase contrast images; **B,D** are same field viewed through epi-fluorescence. **A,B** Ethanol control; *arrow* indicates mitotic cell. **C,D** 1,25(OH)₂D₃ treated; *arrows* indicate apoptotic cells

1,25(OH)₂D₃ treatment. Immunoblots of cell lysates indicate an increase in the mature form of cathepsin B, which migrates at approximately 30 kDa. On immunofluorescence, cathepsin B is indeed localized to intracellular vesicles and apoptotic bodies in 1,25(OH)₂D₃-treated cultures. Parallel northern blotting experiments indicate that the steady state mRNA level of cathepsin B is increased by 1,25(OH)₂D₃ treatment as well (J. Welsh and M. Tenniswood, unpublished observations). These results suggest that ACD in MCF-7 cells treated with 1,25(OH)₂D₃ is associated with enhanced lysosomal activity, consistent with early work describing the induction of lysosomal enzymes during mammary tumor or mammary gland regression in vivo (Lanzerotti and Gullino 1972; Cho-Chung and Gullino 1973; Guenette et al. 1994).

Consistent with previous reports in MCF-7 cells undergoing ACD in response to antiestrogens or serum deprivation (Oberhammer et al. 1993; Warri et al. 1993), we have not detected DNA fragmentation using conventional agarose gel electrophoresis (not shown) despite clear morphological and biochemical evidence of ACD in 1,25(OH)$_2$D$_3$-treated MCF-7 cells. This could reflect the low numbers of cells undergoing ACD at any one time (estimated at about 5% in treated cultures) or the rapid secondary necrosis of apoptotic cells following release from the monolayer. Alternatively, recent evidence indicates that DNA fragmentation into the characteristic 200-bp ladder is a late event in the apoptotic cascade which can be dissociated from the morphological changes (Cohen et al. 1992).

10.2.1.3 Clusterin/TRPM-2

Increased clusterin (TRPM-2) gene expression is associated with apoptotic regression in a variety of cell types; however, limited data exist on expression and subcellular localization of the clusterin protein. On immunofluorescence examination, control MCF-7 cells exhibit a low but detectable level of clusterin expression (Fig. 6). Clusterin immunoreactivity predominates in cytosol, with a punctate distribution. In cells treated with 1,25(OH)$_2$D$_3$ for 48 h, dramatic increases in clusterin intensity are evident, particularly in areas around cells exhibiting morphological evidence of apoptosis. Immunoreactivity is sharply demarcated in apoptotic bodies and secretory vesicles. In both ethanol and 1,25(OH)$_2$D$_3$-treated cultures, large secretory vesicles which do not contain clusterin are present in cells which do not display apoptotic morphology, indicating that the secretory vesicles themselves are not markers of ACD in these cells. However, secretory vesicles of cells exhibiting apoptotic morphology are consistently positive for clusterin.

In western blots of MCF-7 cells, the clusterin antibody specifically recognizes several bands corresponding to the unprocessed and mature forms of clusterin (Fig. 7). The band detected at approximately 68 kDa on sodium dodecyl sulfate-polyacrylamide gel electrophoresis (SDS-PAGE) corresponds to the molecular weight of clusterin endogenously present in MDCK cells or stably expressed in BHK cells and has been shown to represent the high mannose form of the protein (Urban et al. 1987; Pilarsky et al. 1993). The mature form, a disulfide-linked heterodimer, is detected as a doublet at approximately 40 kDa. Clusterin

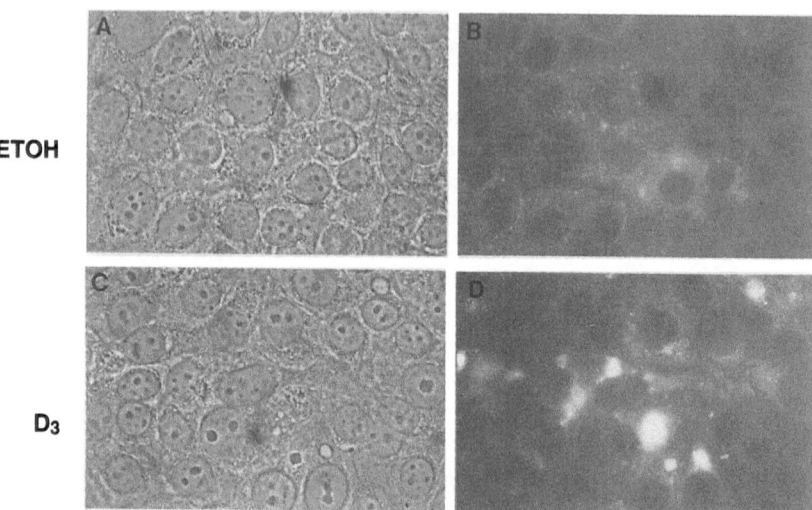

ETOH

D₃

Fig. 6A–D. Immunofluorescence of clusterin/TRPM-2 in MCF-7 cells treated with ethanol or 1,25(OH)$_2$D$_3$. MCF-7 cells grown on coverslips were treated with 100 n*M* 1,25(OH)$_2$D$_3$ or vehicle for 48 h, fixed and processed for immunofluorescence using a monoclonal antibody to clusterin/TRPM-2 and fluorescein isothiocyanate (FITC)-conjugated anti-mouse secondary antibody. Cells were viewed under either phase contrast (**A,C**) or epi-fluorescence (**B,D**). **A,B** ethanol control; **C,D** 1,25(OH)$_2$D$_3$ treated

immunoreactivity is also detected in the media of MCF-7 cells (not shown). Although pulse chase experiments will be necessary to further characterize the processing of clusterin in MCF-7 cells, our data suggest that clusterin is expressed, glycosylated, proteolytically processed, and secreted in MCF-7 cells in a manner similar to that reported for MDCK cells.

Subcellular fractionation indicates that clusterin is present in cytosol, membrane, and nuclear fractions. The recovery of clusterin in the membrane fraction is expected, since it is known to associate with membrane following secretion. The presence of clusterin in the nuclear fraction may represent association with the nuclear membrane. Alternatively, since the endoplasmic reticulum can fractionate with the nuclear membrane, this may represent clusterin undergoing processing (Urban et al.

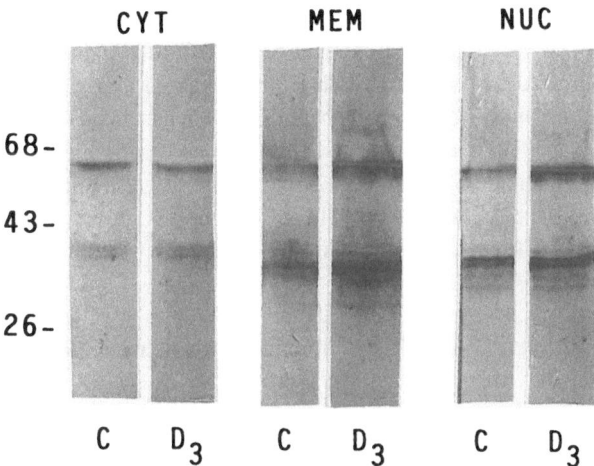

Fig. 7. Western blot of clusterin/TRPM-2 in subcellular fractions of MCF-7 cells treated with ethanol (as controls, *C*) or 1,25(OH)$_2$D$_3$ (D$_3$). Cytosolic (*CYT*), nuclear (*NUC*), and membrane (*MEM*) fractions derived from MCF-7 cells treated with ethanol or 1,25(OH)$_2$D$_3$ for 48 h were separated on 10 % sodium dodecyl sulfate-polyacrylamide gel electrophoresis (SDS-PAGE), transferred to nitrocellulose, and immunoblotted with a monoclonal antibody to clusterin/TRPM-2. Alkaline phosphatase-conjugated secondary antibody was detected with the Protoblot system from Promega (Madison, WI, USA)

1987). In 1,25(OH)$_2$D$_3$-treated cells, an increased expression, primarily of the unprocessed form, is detected in both membrane and nuclear fractions. Time course studies indicated that clusterin expression is increased within 24 h of 1,25(OH)$_2$D$_3$ treatment, coincident with the first morphological evidence of ACD (not shown). Enhanced expression of the steady state mRNA for TRPM-2 is also observed in 1,25(OH)$_2$D$_3$-treated cells. Thus, immunofluorescence, immunoblotting, and northern analysis suggest that more clusterin is synthesized in 1,25(OH)$_2$D$_3$-treated cells. Studies are in progress to determine whether 1,25(OH)$_2$D$_3$ affects the processing of clusterin in MCF-7 cells during the apoptotic process. Although the function of clusterin is incompletely understood, its association with membrane and nuclear fractions in MCF-7 cells suggests a role in the remodeling of membranes during apoptotic body formation.

Taken together, our morphological and biochemical data suggest that 1,25(OH)$_2$D$_3$ or EB1089 treatment of MCF-7 cells induces a specific subtype of ACD (type II) which is characterized by cytoplasmic and chromatin condensation, reorganization of nuclear matrix proteins, lysosomal activation, and enhanced clusterin expression in the absence of DNA fragmentation. Studies are currently in progress to determine the nature of the regression process in MCF-7 tumors grown as xenografts in nude mice treated with vitamin D compounds.

10.2.2 Effects of Vitamin D Compounds on Estrogen Receptor and TGF-β

MCF-7 cells are estrogen dependent and undergo ACD upon estrogen withdrawal or treatment with antiestrogens (Kyrpianou et al. 1991; Warri et al. 1993). In addition, the effects of 1,25(OH)$_2$D$_3$ and EB1089 on MCF-7 cell cycle kinetics described above are strikingly similar to those elicited by antiestrogens (Osborne et al. 1985). These similarities suggest that vitamin D compounds could elicit their effects via disruption of estrogen receptor signaling. In support of this suggestion, it has recently been reported that 1,25(OH)$_2$D$_3$ attenuates induction of the estrogen regulated gene pS2 by estradiol in MCF-7 cells (Demirpence et al. 1994). We thus compared the effects of 1,25(OH)$_2$D$_3$ and EB1089 on the estrogen receptor in MCF-7 cells, using ligand binding and immunoblotting techniques. Both 1,25(OH)$_2$D$_3$ and EB1089 decrease [^{3}H]estradiol binding approximately 30% within 24 h of exposure. The decrease in estradiol binding is associated with a decreased amount of receptor detected on immunoblots (not shown). The observed down-regulation of estrogen receptor within 24 h of treatment precedes the inhibitory effect of 1,25(OH)$_2$D$_3$ on growth indices (Fig. 2), suggesting that induction of ACD may, at least partially, involve disruption of estrogen-dependent signals.

One of the downstream effects of estrogen signaling in MCF-7 cells is suppression of TGF-β expression (Knabbe et al. 1987; Jeng and Jordan 1991). Induction of ACD of breast cancer cells upon estrogen ablation or treatment with chemotherapeutic agents is associated with enhanced expression of TGF-β mRNA (Armstrong et al. 1992; Kyprianou et al. 1991; Warri et al. 1993), and TGF-β has been shown to induce

Table 2. Effect of neutralizing antibody to TGF-β on MCF-7 cell number following treatment with 1,25(OH)$_2$D$_3$ or EB1089[a]

TGF-β neutralizing antibody	Cell number ($\times 10^4$/ml, mean ± SEM)		
	Ethanol	1,25(OH)$_2$D$_3$	EB1089
Absent	159 ± 22	73 ± 4[*]	97 ± 10[*]
Present	202 ± 43	210 ± 25[**]	188 ± 35[**]

[a]Cells were treated with 100 nM 1,25(OH)$_2$D$_3$, 1 nM EB1089 (± antibody) for 48 h. Significant effect of vitamin D compound (*) or neutralizing antibody (**).

ACD in a variety of cell types (Martikainen et al. 1990; Rotello et al. 1991; Oberhammer et al. 1991). We have observed that treatment of MCF-7 cells with 2 ng/ml TGF-β$_1$ for 48 h induces morphological changes similar to those induced in response to vitamin D analogs (not shown). We thus hypothesize that 1,25(OH)$_2$D$_3$ may induce apoptosis via enhancement of TGF-β secretion and/or activation. To determine whether the effect of 1,25(OH)$_2$D$_3$ in MCF-7 cells could be linked to TGF-β, we first examined the effect of a neutralizing antibody to TGF-β. As indicated in Table 2, this antibody, which neutralizes all isoforms of TGF-β, completely blunts the effects of 1,25(OH)$_2$D$_3$ or EB1089 on cell number. Parallel studies indicate that coincubation of the TGF-β neutralizing antibody prevents the characteristic reorganization of nuclear matrix proteins in MCF-7 cells treated with 1,25(OH)$_2$D$_3$ or EB1089 (M. Simboli-Campbell and J. Welsh, manuscript submitted). These data suggest that secretion and/or activation of TGF-β by MCF-7 cells is enhanced following treatment with 1,25(OH)$_2$D$_3$ or EB1089. This suggestion is supported by additional observations from our laboratory indicating that both vitamin D compounds increase the ratio of unphosphorylated to phosphorylated Rb protein in MCF-7 cells, an effect which mimics that of TGF-β. These results indicate that TGF-β likely participates in the process of ACD, rather than being secreted by dying cells as a consequence of ACD. Consistent with these observations, 1,25(OH)$_2$D$_3$ modulates both expression and bioactivity of TGF-β in keratinocytes and induces dephosphorylation of Rb protein in keratinocytes (Kim et al. 1992; Kobayashi et al. 1993). Further, keratinocyte growth inhibition in response to 1,25(OH)$_2$D$_3$ is partially reversed in the presence of neutralizing antibodies to TGF-β (Kim et al. 1992). Further work will be necessary to determine whether vitamin D compounds modulate TGF-β directly or whether this effect is primarily related to downregulation of estrogen signaling. Since 1,25(OH)₂D₃ inhibits growth of estrogen receptor-negative breast cancer cells (Chouvet et al. 1986, Abe et al. 1992), it is likely that vitamin D compounds exert effects via multiple mechanisms, some of which are unrelated to estrogen-mediated events. Significantly, since the vitamin D receptor and the estrogen receptor do not always colocalize in human breast tumors, combination therapies employing antiestrogens or anti-progestins and vitamin D compounds may enhance regression of tumors containing both estrogen receptor positive- and estrogen receptor-negative cells.

The potent hypercalcemic effect of 1,25(OH)₂D₃ precludes its use in vivo to induce tumor regression, thus novel vitamin D analogs such as EB1089 have been developed which lack calcemic effects. As we report here, EB1089 (which lacks calcemic effects in vivo) induces ACD in MCF-7 cells with a potency 100 times that of 1,25(OH)₂D₃, emphasizing its potential for clinical use. However, the newly recognized actions of vitamin D compounds in induction of ACD raise the possibility that these compounds might induce ACD in other target tissues for vitamin D. Intestine and kidney are of particular interest, since these tissues exhibit the highest level of vitamin D receptor (VDR) in vivo and thus would be most susceptible to such side effects of vitamin D analogs. With respect to target tissue specificity of 1,25(OH)₂D₃'s effects in vitro, we have studied the effects of 1,25(OH)₂D₃ on normal renal epithelial cells (MDBK) and intestinal crypt cells (IEC-6), two cell lines which express vitamin D receptors and known vitamin D-regulated proteins. Although 1,25(OH)₂D₃ inhibits proliferation and induces vitamin D-dependent proteins in both MDBK and IEC-6 cells, no evidence of ACD is observed even when cells are treated with 500 n*M* 1,25(OH)₂D₃ for up to 72 h (unpublished data). We thus hypothesize that coupling between 1,25(OH)₂D₃, the VDR and the apoptotic pathway is cell type specfic. The underlying basis for this specificity is unclear, but we believe it may be related to the transformed phenotype. Studies are in progress to assess indices of ACD in intestine and kidney from nude mice receiving therapeutic doses of 1,25(OH)₂D₃ to induce regression in MCF-7 cells grown as xenografts. If, as our hypothesis predicts, vitamin D compounds specifically induce ACD in tumor cells while sparing normal cells expressing the VDR, vitamin D analogs will offer a clear advantage over existing therapies for human breast cancer.

10.3 Conclusion

Our studies clearly demonstrate that the initial growth inhibitory effect of 1,25(OH)$_2$D$_3$ on MCF-7 human breast cancer cells results from induction of ACD. This conclusion is based on temporal studies which indicate that significant decreases in cell number in 1,25(OH)$_2$D$_3$-treated cells precede any significant effect of 1,25(OH)$_2$D$_3$ on cell proliferation. Morphological and biochemical criteria demonstrate several characteristics of ACD in 1,25(OH)$_2$D$_3$-treated cells, including condensation of nuclear matrix proteins and chromatin, apoptotic body formation, and increased expression of clusterin/TRPM-2. Similar effects on induction of ACD were observed following treatment with EB1089, a vitamin D analog developed by LEO pharmaceuticals which is undergoing testing as a chemotherapeutic agent for breast cancer.

Both 1,25(OH)$_2$D$_3$ and EB1089 downregulate the estrogen receptor in MCF-7 cells, and the effects of both are blunted in the presence of a neutralizing antibody to TGF-β. However, since 1,25(OH)$_2$D$_3$ also inhibits growth of breast cancer cells lacking the estrogen receptor, it is likely that at least some of the effects of 1,25(OH)$_2$D$_3$ are independent of estrogen signaling. If so, more complete tumor regression may be achieved by combinations of vitamin D compounds and antiestrogens. In vitro studies indicate that vitamin D compounds inhibit proliferation but do not induce ACD in normal renal or intestinal epithelial cells, suggesting that the coupling between the VDR and the cell death pathway is cell type specific and may be related to the transformed phenotype. Our studies identify 1,25(OH)$_2$D$_3$ as a coordinate regulator of both proliferation and ACD in breast cancer cells and emphasize the importance of assessing the efficacy of novel vitamin D analogs in inducing the cell death pathway during tumor regression.

Acknowledgements. We would like to thank Dr. M. Wilson, Dr. J. Mort, and Matritech, Inc. for generously supplying the clusterin, cathepsin B, and NUMA antibodies, respectively. The supply of EB1089 from Dr. L. Binderup,

Leo Pharmaceuticals, was greatly appreciated. We would like to acknowledge the expert technical help of Kim Wong with photography and Tim Welsh with cell culture. These studies were supported by a grant to J. Welsh from The American Institute for Cancer Research.

References

Abe J, Nakano T, Nishii Y, Matsumoto T, Ogata E, Ikeda K (1992) A novel vitamin D3 analog, 22-oxa-1,25 dihydroxyvitamin D3, inhibits the growth of human breast cancer in vitro and in vivo without causing hypercalcemia. Endocrinology 129:832–837

Armstrong DK, Isaacs JT, Ottaviano YL, Davidson NE (1992) Programmed cell death in an estrogen independent human breast cancer cell line, MDA-MB-468. Cancer Res 52:3418–3424

Berger U, Wilson P, McClelland RA, Colston K, Haussler MR, Pike W, Coombes RC (1987) Immunocytochemical detection of 1,25-dihydroxyvitamin D3 receptor in breast cancer. Cancer Res 47:6793–6799

Bursch W, Paffe S, Putz B, Barthel G, Schulte-Hermann R (1990) Determination of the length of the histological stages of apoptosis in normal liver and altered hepatic foci of rats. Carcinogenesis 11:847–853

Cho-Chung YS, Gullino PM (1973) Mammary tumor regression V. Role of acid ribonuclease and cathepsin. J Biol Chem 248:4743–4749

Chouvet C, Berger U, Coombes RC (1986) 1,25 Dihydroxyvitamin D3 inhibitory effect on the growth of two human breast cancer cell lines (MCF-7, BT-20). J Steroid Biochem 24:373–376

Cohen GM, Sun XM, Snowden RT, Dinsdale D, Skilleter DN (1992) Key morphological features of apoptosis may occur in the absence of internucleosomal DNA fragmentation. Biochem J 286:331–334

Colston K, Berger U, Wilson P, Hadcocks L, Naeem I, Earl HM, Coombes RC (1988) Mammary gland 1,25-dihydroxyvitamin D3 receptor content during pregnancy and lactation. Mol Cell Endocrinol 60:15–22

Colston K, Chander SK, Mackay AG, Coombes RC (1992) Effects of synthetic vitamin D analogs on breast cancer cell proliferation in vivo and in vitro. Biochem Pharmacol 44:693–702

Demirpence E, Balaguer P, Trousse F, Nicolas JC, Pons M, Gagner D (1994) Antiestrogenic effects of all trans retinoic acid and 1,25 dihydroxyvitamin D in breast cancer cells occur at the estrogen response element level but through different molecular mechanisms. Cancer Res 54:1458–1464

Eisman J, Sutherland RL, McMenemy ML, Fragonas JC, Musgrove EA, Pang G (1989) Effects of 1,25-dihydroxyvitamin D3 on cell cycle kinetics of T47D human breast cancer cells. J Cell Physiol 138:611–616

Frampton RJ, Omond SA, Eisman JA (1983) Inhibition of human cancer cell growth by 1,25-dihydroxyvitamin D$_3$ metabolites. Cancer Res 43:4443–4447

Guenette RS, Corbeil H, Leger J, Wong K, Mezl V, Mooibroek M, Tenniswood M (1994) Induction of gene expression during involution of the lactating mammary gland of the rat. J Mol Endocrinol 12:47–60

Jeng MH, Jordan VC (1991) Growth stimulation and differential regulation of transforming growth factor-beta 1 (TGFB1), TGFB2 and TGFB3 messenger RNA levels by norethidrone in MCF-7 human breast cancer cells. Mol Endocrinol 5:1120-1128

Kim H, Abdelkader N, Katz M, McLane JA (1992) 1,25-Dihydroxyvitamin D$_3$ enhances antiproliferative effect and transcription of TGF-beta on human keratinocytes in culture. J Cell Physiol 151:579–587

Knabbe C, Lippman ME, Wakefield LM, Flanders KC, Kasid A, Derynck R, Dickson RB (1987) Evidence that transforming growth factor-beta is a hormonally regulated negative growth factor in human breast cancer cells. Cell 48:417-428

Kobayashi T, Hashimoto K, Yoshikawa K (1993) Growth inhibition of human keratinocytes by 1,25-Dihydroxyvitamin D$_3$ is linked to dephosphorylation of retinoblastoma gene product. Biochem Biophys Res Commun 196:487–493

Kyprianou N, English H, Davidson N, Isaacs J (1991) Programmed cell death during regression of the MCF-7 human breast cancer following estrogen ablation. Cancer Res 51:162–166

Lanzerotti RH, Gullino PM (1972) Activities and quantities of lysosomal enzymes during mammary tumor regression. Cancer Res 32:2679–2685

Martikainen P, Kyprianou N, Isaacs JT (1990) Effect of transforming growth factor β1 on proliferation and death of rat prostatic cells. Endocrinology 127:2963–2968

Miller TE, Beausang LA, Meneghini N, Lidgard G (1994) Cell death and nuclear matrix proteins. In: Tomei LD, Cope FO (eds) Apoptosis II: the molecular basis of apoptosis in disease. Cold Spring Harbor Laboratory Press, Cold Spring Harbor, pp 357–376

Oberhammer F, Bursch W, Parzefall W, Breit P, Erber E, Stadler M, Schulte-Hermann R (1991) Effect of transforming growth factor beta on cell death of cultured rat hepatocytes. Cancer Res 51:2478–2485

Oberhammer F, Wilson JW, Dive C, Morris ID, Hickman JA, Wakeling AE, Walker PR, Sikorska M (1993) Apoptotic death in epithelial cells: cleavage of DNA to 300 and/or 50 kb fragments prior to or in the absence of internucleosomal degradation. EMBO J 12:3679–3684

Osborne CK, Boldt DH, Clark GM, Trent JM (1985) Effects of tamoxifen on human breast cancer cell cycle kinetics: accumulation of cells in early G$_1$ phase. Cancer Res 43:3583–3585

Pilarsky C, Haase W, Koch-Brandt C (1993) Stable expression of gp80 (TRPM-2, clusterin), a secretory protein implicated in programmed cell death, in transfected BHK-21 cells. Biochim Biophys Acta 1179:306–310

Rotello RJ, Lieberman RC, Purchio AF, Gerschenson LE (1991) Coordinated regulation of apoptosis and cell proliferation by transforming growth factor beta 1 in cultured uterine epithelial cells. Proc Natl Acad Sci USA 88:3412–3415

Sahota SS, Edgar AJ, Colston K, Coombes RC (1991) Analysis of 1,25-dihydroxyvitamin D3 mRNA in human breast cancer tissues by polymerase chain reaction and in rat mammary tumours by Northern blots. In: Norman AW, Bouillon R, Thomasset M (eds) Vitamin D: gene regulation, structure-function analysis and clinical application. Proceedings of the 8th workshop on vitamin D, 5-10 July 1991, Paris, France. De Gruyter, Berlin, pp 459-460

Tenniswood M, Guenette RS, Lakins J, Mooibroek M, Wong P, Welsh JE (1992) Active cell death in hormone dependent tissues. Cancer Metastasis Rev 11:197–220

Tenniswood M, Taillefer D, Lakins J, Guenette RS, Mooibroek M, Daehlin L, Welsh JE (1994) Control of gene expression during apoptosis in hormone-dependent tissues. In: Tomei LD, Cope FO (eds) Apoptosis II: the molecular basis of apoptosis in disease. Cold Spring Harbor Laboratory Press, Cold Spring Harbor, pp 283–311

Urban J, Parczyk K, Leutz A, Kayne M, Kondor-Kock C (1987) Constitutive apical secretion of an 80-kD sulfated glycoprotein complex in the polarized epithelial Madin Darby canine kidney cell line. J Cell Biol 105:2735–2743

Walker PR, Kwast-Welfeld J, Gourdeau H, Leblanc J, Neugebauer W, Sikorska M (1993) Relationship between apoptosis and the cell cycle in lymphocytes: roles of protein kinase C, tyrosine phosphorylation and AP1. Exp Cell Res 207:142–151

Warri AM, Huovinen RL, Laine AM, Martikainen PM, Harkonen PL (1993) Apoptosis in toremifene-induced growth inhibition of human breast cancer cells in vivo and in vitro. J Natl Cancer Inst 85:1412–1418

Williams GT (1991) Programmed cell death: apoptosis and oncogenesis. Cell 65:1097–1098

Yang CH, Lambie EJ, Snyder M (1991) NUMA: an unusually long coiled-coil related protein in mammalian nucleus. J Cell Biol 116:1303–1317

11 Vitamin D_3 Derivatives and Breast Cancer

K. W. Colston, A. G. Mackay and S. Y. James

11.1 Metabolism of Vitamin D

Vitamin D_3 is formed in the skin from the precursor molecule 7-dehydrocholesterol in response to exposure to solar ultraviolet irradiation (UVR). When 7-dehydrocholesterol is irradiated with ultraviolet light, the bond between carbons 9 and 10 in ring B is broken and ring A rotates through 180° around the single bond between carbons 6 and 7 to form vitamin D_3 or cholecalciferol (Fig. 1). There are few foodstuffs which contain appreciable amounts of vitamin D – the chief sources are deep sea fish, milk fat and egg yolk. Exposure to UV light is the main determinant of vitamin D status. Intake of vitamin D from an unsupplemented diet is minimal and does not contribute significantly to vitamin D status. Even in high latitudes most (~ 90%) vitamin D is

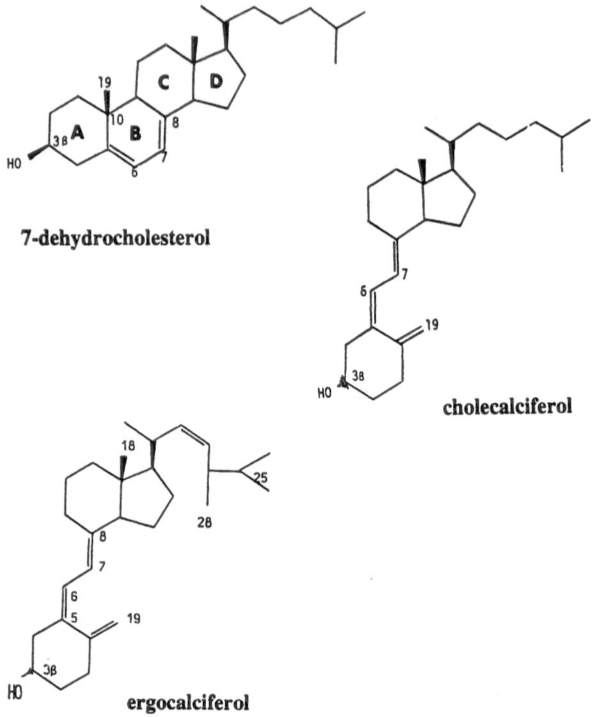

Fig. 1. Structures of 7-dehydrocholesterol, cholecalcifierol (vitamin D₃) and ergocalciferol (vitamin D₂)

synthesized in the skin with ~10% derived from dietary sources. Figure 1 also shows the structure of vitamin D_2 or ergocalciferol which occurs rarely in nature, but its metabolism and mechanism of action are essentially the same as those of vitamin D_3.

Vitamin D_3 formed in the skin or consumed in the diet is transported to the liver where it is hydroxylated to 25-hydroxyvitamin D_3, the major circulating form (Fig. 2). The 25-hydroxylase enzyme is virtually unregulated so that any vitamin D_3 presented to the liver is converted to 25-hydroxyvitamin D_3. In blood, this metabolite is present at concentrations on the order of nanograms per millilitre and is regarded as a useful index of vitamin D status. Small amounts of 25-hydroxyvitamin D_3 are subsequently converted in the kidney to 1-α-25-dihydroxyvitamin D_3

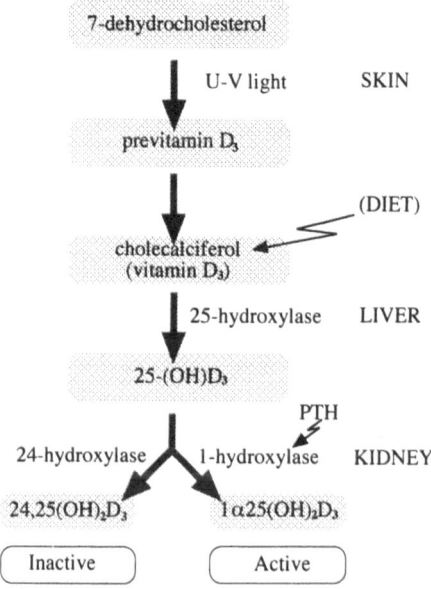

Fig. 2. Metabolism of vitamin D

[1,25(OH)$_2$D$_3$], the active hormonal form which is present in the circulation in picogram quantities per millilitre. The activity of the renal 1-α-hydroxylase enzyme is regulated by a number of factors including the circulating concentration of parathyroid hormone. Levels of 1,25(OH)$_2$D$_3$ remain fairly constant in the face of substantial changes in vitamin D$_3$ synthesis or intake.

11.2 Mechanism of Action of 1,25-Dihydroxyvitamin D₃

The mechanism of action of 1,25(OH)$_2$D$_3$ is analogous to that of other steroid hormones. In serum, the hormone circulates bound to a specific transport protein. The free fraction diffuses into target cells and interacts with its specific intracellular receptor, enabling the receptor protein to bind to specific DNA sequences in the promotor region of target genes, which then leads to changes in gene transcription. In this way new messenger RNAs are produced which code for bioactive proteins. These include proteins involved in intestinal calcium transport, bone matrix proteins, the enzymes alkaline phosphatase and 25-hydroxyvitamin D-24-hydroxylase and also as yet unidentified proteins which are involved with cell differentiation and suppression of the malignant phenotype.

In recent years considerable effort has been exerted by a number of groups in characterizing the vitamin D receptor protein which is present in trace amounts in target cells. The cloning of the structural gene for the vitamin D receptor (VDR) has been described (McDonnell et al. 1987). From the deduced amino acid sequence, it is clear that VDR displays considerable sequence homology with other hormone receptors which belong to the steroid, retinoic acid and thyroid hormone receptor super-family (Evans 1988). These proteins share similarities in their structural organization, comprising functional domains. Of importance are the DNA recognition domain near the N-terminus where considerable se-quence homology exists between the receptor protein family members, and a C-terminal ligand binding domain where homology is less marked. The DNA binding domain of the various receptor proteins are envisaged as comprising two finger-like structures, each folded about a single zinc atom (the "zinc fingers"). This region of the receptor protein contains sequences which appear to recognize specific hormone-respon-sive elements located in genomic DNA adjacent to target genes.

11.3 Actions of 1,25-Dihydroxyvitamin D3 on Intestine and Bone

The major role of $1,25(OH)_2D_3$, acting via its receptor protein, is to maintain calcium homeostasis by its actions on intestinal calcium ab-sorption and bone. In the intestine, $1,25(OH)_2D_3$ increases active cal-cium transport in the small intestine. The molecular mechanisms in-volved are not completely understood but seem to involve the de novo production of calbindin 9K, a calcium-binding protein. Actions of $1,25(OH)_2D_3$ on bone are complex. Stimulatory effects of the hormone have been shown on bone resorption and $1,25(OH)_2D_3$ promotes dif-ferentiation of precursor cells into mature osteoclasts. Clear stimulatory effects on osteoblastic function have also been documented. Adequate intake of vitamin D and of calcium is essential for skeletal health.

Vitamin D deficiency gives rise to osteomalacia in adults and to rickets in children. Privational rickets or osteomalacia may develop with inadequate exposure to sunlight together with insufficient dietary intake. Vitamin D deficiency is not uncommon in the British Asian population although the prevalence of privational rickets and osteomalacia in the community has declined in the last decade. Since vitamin D_3 is produced in the skin in response to solar UVR, a seasonal variation of 25-hydroxyvitamin D in plasma is seen with peak levels observed in the late summer months and trough levels in late winter to early spring. Both peak and trough levels have been found to be significantly lower in healthy Asian subjects living in London than in white subjects (Finch et al. 1992). Studies on the aetiology of vitamin D deficiency in British Asians have revealed that limited ultraviotet light exposure is necessary but insufficient by itself to induce the biochemical, radiological and clinical expression of privational rickets and osteomalacia. In a diet unfortified with vitamin D, the major food class risk factors for development of osteomalacia are high intakes of high extraction and wholemeal cereals and low or absent meat intake (Dunnigan 1992).

11.4 Actions of 1,25-Dihydroxyvitamin D3 on Cancer Cells In Vitro

In recent years it has become clear that $1,25(OH)_2D_3$ has actions in other tissues apparently unrelated to calcium handling. Receptors are not confined to intestine and bone. Biochemical, autoradiographic and immunocytochemical studies (Berger et al. 1988) have revealed numerous potential target organs for the hormone, including kidney, brain, liver, skin, reproductive tissues, hormone-secreting glands and certain cells of the immune system (Table 1). The function of $1,25(OH)_2D_3$ in many of these tissues remains to be elucidated and many established human cancer cell lines are VDR positive. In 1981 it was first demonstrated that $1,25(OH)_2D_3$ inhibits the growth of cancer cells in vitro (Colston et al. 1981). These first experiments were performed in amelanotic melanoma cells and demonstrated that $1,25(OH)_2D_3$ at concentrations in the nanomolar range significantly inhibits cell proliferation. Later that same year the observation was made that $1,25(OH)_2D_3$ can promote the differentiation of leukaemic

Table 1. Localization of vitamin D receptor in normal human tissue (adapted from Berger et al. 1988)

Epithelial tissue	
Duodenum[a]	++
Jejunum	++
Colon	+++
Kidney (tubular cells)	++/+++
Kidney (glomeruli)	–
Liver (hepatocytes)	+/++
Liver (biliary ducts)	+[b]
Skin epidermis	++
Skin appendages	++
Breast epithelium (normal)	++
Endometrium	+[b]
Tyroid (nodular goiter)	++/+++
Adrenal (cortex)	+/++
Oesophagus	++
Stomach	+/++[b]
Lung (bronchial epithelial cells)	+/++[b]
Lung (alveolar cells)	–
Urethral mucosa (adjacent prostatic glands)	+
Prostatic glands	+[b]
Nonepithelial tissue	
Vessel walls	+[c]
Fibrocytes	+
Smooth muscle[c]	+[c]
Fetal bone	+
Striated muscle (pectoral)	–

Nuclear immunostaining varied from weakly positive (+) to moderately positive (++), to strong positive (+++).
[a]Gradation of staining: crypts +++, villus tips o/+.
[b]Some tissues showed a more focal reaction.
[c]Normal and cytoplasmic immunostaining.
(–) No staining.

cells (Abe et al. 1981). Such observations prompted the suggestion that $1,25(OH)_2D_3$ might be of use in treating hyperproliferative disorders such as psoriasis and leukaemia, but a major drawback to considering conventional vitamin D compounds as therapeutic agents is their potent calcaemic activity, leading to hypercalciuria and hypercalcaemia at doses of more than a few micrograms per day.

11.5 Vitamin D and Cancer

11.5.1 Epidemiological Studies

At around the same time, epidemiological studies were pointing to a relationship between vitamin D and cancer. It was suggested in 1980 that calcium and vitamin D may reduce the risk of colon cancer, the assertion being derived from the geographic epidemiology of death rates of colon cancer which tend to increase with increasing latitude and decreasing sunlight intensity (Garland and Garland 1980). Dietary calcium is also associated with reduced incidence of colon cancer (Slattery et al. 1988) and more recent results demonstrated an inverse relationship between serum 25-hydroxyvitamin D and incidence of colonic cancer (Garland et al. 1989). A geographic variation in breast cancer mortality in the US has been noted, suggesting that synthesis of vitamin D from sunlight exposure may be associated with low risk for fatal breast cancer (Garland et al. 1990). However, the correlation between vitamin D and breast cancer has been challenged by Simard and collegues (1991) who noted similar vitamin D intakes in breast cancer patients and control subjects. In addition, Schwartz and Hulka (1990) have suggested that mortality from prostate cancer might also be linked to vitamin D because of a geographic relationship between mortality from this malignancy and UV light intensity similar to that for breast and colon cancer.

11.5.2 Laboratory Studies

Studies from several laboratories including our own have demonstrated that a large proportion of breast tumour biopsy specimens contain VDR although there is no apparent correlation between VDR positivity and oestrogen receptor status (Berger et al. 1987). In order to determine the possible prognostic significance of VDR in breast cancer, breast tumour biopsy specimens from 136 patients with primary carcinoma of the breast were examined for the presence of the receptor by an immunocytochemical technique. The results suggest that VDR status may be positively related to disease-free survival time; the patients with VDR negative tumours relapsed significantly earlier than those with VDR-positive tumours (Colston et al. 1989). These findings may indicate that circulating $1,25(OH)_2D_3$, acting via its intracellular receptor, may influence tumour activity.

In animal studies, administration of $1,25(OH)_2D_3$ and its analogue alfacalcidol [which is converted to $1,25(OH)_2D_3$ in vivo] prolonged the survival time of mice inoculated with myeloid leukaemia cells (Honma et al. 1983). In another study, immunodeficient mice bearing tumour xenografts of colon carcinomas and melanomas were treated with $1,25(OH)_2D_3$ and inhibition of tumour growth was reported (Eisman et al. 1987). However, the animals were maintained on a low calcium diet to limit intestinal absorption of calcium. Our own studies have utilized an animal model of hormone-dependent breast cancer to evaluate antitumour effects of vitamin D derivatives. In this model, mammary tumours are induced in adult female rats with the carcinogen nitrosomethylurea (NMU). Tumours are oestrogen dependent and regress when the animals are ovariectomized. Our original studies demonstrated that alfacalcidol given intraperitoneally significantly inhibited the progression of these tumours, but with increasing doses marked hypercalcaemia developed, even when the animals were maintained on a low calcium diet (Colston et al. 1992). These laboratory studies highlight the drawback to considering conventional vitamin D metabolites as therapeutic agents because of their potent calcaemic activity leading to the risk of hypercalcaemia, hypercalciuria and development of nephrocalcinosis.

There have been few clinical trials of vitamin D in cancer. A trial of alfacalcidol in patients with low-grade non-Hodgkins lymphoma showed response in a proportion of patients (Cunningham et al. 1985). This compound was also assessed in a trial of patients with myelodysplasia. A transient improvement in peripheral blood counts was seen but half of the patients developed hypercalcaemia (Koeffler et al. 1985). A second report from Kelsey and his colleagues (Kelsey et al. 1992) showed sustained haematological response in six patients treated with high-dose alfacalcidol. In this study patients were given dietary advice to limit calcium intake. Four patients developed hypercalcaemia and received treatment with the bisphosphonate APD to limit bone resorption.

11.6 New Analogues of Vitamin D

Interest in our group has centred on the therapeutic potential of new vitamin D analogues whose profile of activity displays enhanced anti-proliferative effects relative to calcaemic activity. Modification of the C17 side chain of the vitamin D molecule seems to be the most effective approach to separate calcaemic from antiproliferative activities. One compound, MC903 (calcipotriol from Leo Pharmaceutical Products and shown in Fig. 3), contains a cyclopropyl substitution in the side chain and appears to be equipotent with $1,25(OH)_2D_3$ in inhibiting the proliferation of MCF-7 breast cancer cells in vitro. In contrast, a 100–200-fold greater dose of MC903 than $1,25(OH)_2D_3$ is required in vivo to give an equivalent effect on serum and urinary calcium in experimental animals (Binderup and Bramm 1988; Colston et al. 1992).

In clinical trials, calcipotriol has been shown to be effective in the treatment of psoriasis (Kragballe 1989). Recently its value in topical therapy in cutaneous nodules in patients with advanced breast cancer has been assessed (Bower et al. 1991). This trial involved evaluation of 19 patients with resistant locally advanced breast cancer or with cutaneous metastases which could be measured bidimensionally. Fourteen patients completed 6 weeks of treatment and of these three showed partial response and one a minimal response of the treated nodule. The presence of VDR in tumour cells (from fine needle aspirates or, where insufficient material was available, in sections of primary tumour) was evaluated by immunocytochemistry. Receptor protein was detected in tumour cells from seven of 13 patients, including all four responders (Table 2). This study introduces a potential new class of endocrine therapy in breast cancer. Calcipotriol has a relatively short half-life in vivo but similar and metabolically more stable compounds could be an approach to systemic therapy. More recently developed vitamin D analogues with reduced calcaemic activity possess potent antiproliferative activity in breast cancer cells in vitro but are longer lived in vivo. We

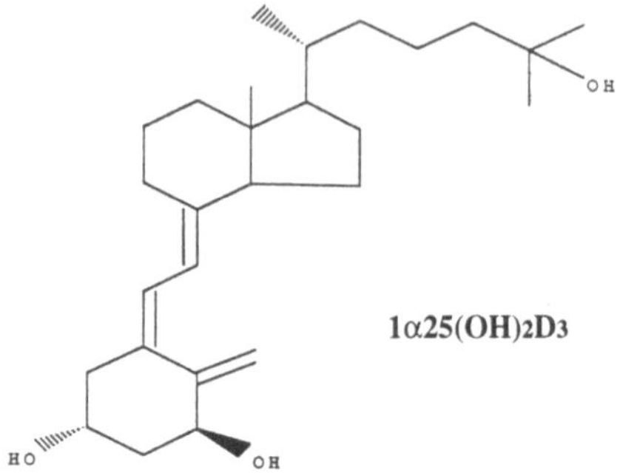

Fig. 3. Structures of 1-α-25-dihydroxyvitamin D₃ and its synthetic analogue calcipotriol (MC903)

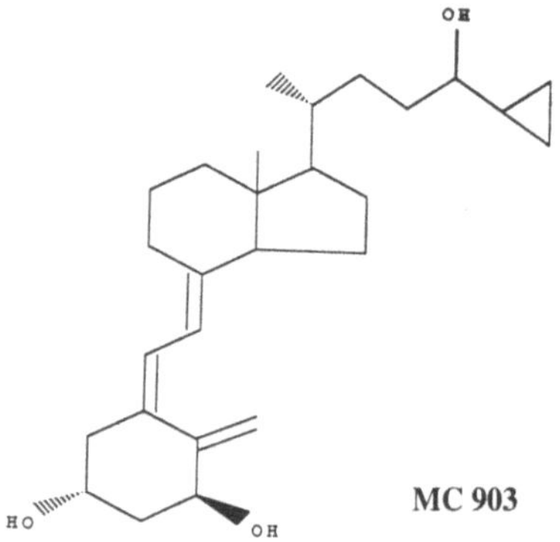

have undertaken preclinical trials of a number of such compounds in order to determine their effectiveness in causing regression of experimental mammary tumours in vivo. Figure 4 shows the structures of some of the compounds which we have examined. All are characterized by modifications in the side chain of the vitamin D molecule. In addition, the 20-epi compounds MC1301, KH1049 and KH1060 display altered stereochemistry in the C20 position (Binderup et al.1991). Figure 4 also shows the concentration of each compound required to produced 50% inhibition of MCF-7 cell proliferation relative to control cultures. Under the culture conditions used, a concentration of $7 \times 10^{-12}M$ KH1060 was required to achieve this compared to $3.5 \times 10^{-9}M$ for

Vitamin D_3 Derivatives and Breast Cancer   211

Side chain	Name	MCF-7 cell growth inhibition IC_{50}	calcaemic activity
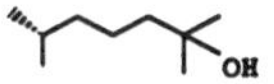	1,25(OH)2D3	3.5×10^{-9}M	100%
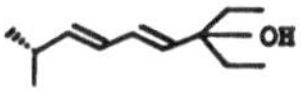	EB1089	1.5×10^{-10}M	40%
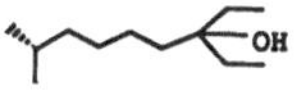	CB966	1.2×10^{-9}M	20%
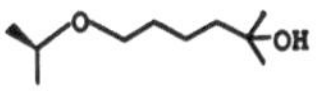	KH1049*	1.5×10^{-11}M	140%
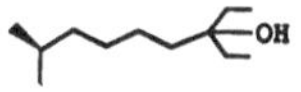	MC1301*	1.8×10^{-10}M	120%
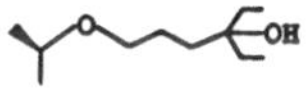	KH1060*	7×10^{-12}M	130%

* 20 - epi compounds

Fig. 4. Structure–activity relationships of vitamin D analogues. MCF-7 cells (2×10^4 cells/ml) were cultured for 7 days in the presence of 1,25(OH)$_2$D$_3$ or vitamin D analogues. All compounds were tested in three separate experiments. Cell proliferation was determined by [³H]thymidine incorporation into trichloracetic acid precipitable material. Calcaemic activity refers to the increase in urinary calcium excretion induced in normal rats by each compound in relation to that produced by 1,25(OH)$_2$D$_3$

1,25(OH)$_2$D$_3$. EB1089 was also considerably more potent than the native hormone. Calcaemic activity refers to the increase in urinary calcium excretion induced in normal rats by each compound in relation to that produced by 1,25(OH)$_2$D$_3$. In this regard, the activity of both CB966 and EB1089 is reduced relative to the native hormone (Binderup et al. 1991; Mathiasen et al. 1993).

Figure 5 demonstrates the effects of these compounds on the growth of established NMU-induced rat mammary tumours. No significant inhibition of tumour progression was seen with the 20-epi compounds KH1060, KH1049 or MC1301 at the doses used but all produced a significant increase in serum calcium concentration (Fig. 5). In contrast, CB966 at 1 µg/kg and EB1089 at 0.5 µg/kg caused a significant inhibi-

Table 2. Characteristics and treatment outcome of assessable patients (adapted from Bower et al. 1991)

Age (years)	Objective response	Planimetry ratio	Control nodules	VDR status skin nodule	VDR status primary tumour
72	PR	0.36	PD	+ve	
65	PR	0.36	PD	+ve	
51	NC	0.9	NC	−ve	
69	NC	0.68	PD	−ve	
55	NC	N/A	N/A[a]	−ve	
70	NC	N/A	N/A[a]	−ve	
64	PR	0.42	NC	+ve	
64	PD	N/A	N/A[a]	−ve	
56	MR	0.73	NC	N/A	+ve
38	PD	0.98	PD	N/A	+ve
75	PD	N/A	N/A[a]	N/A	N/A
88	PD	1.13	PD	N/A	+ve
73	NC	1.5	NC	+ve	
64	PD	3.9	NC	−ve	

[a]In four patients only a single nodule was present and no control was used.
MR, minimal response; PR, partial response; PD, progressive disease; N/A, not available/inadequate sample; NC, no change.

tion of tumour progression, in the case of EB1089 in the absence of a significant rise in serum calcium concentration. The same dose of $1,25(OH)_2D_3$ had no significant effect on tumour growth, but the treated animals developed hypercalcaemia. Thus EB1089 displays enhanced

Fig. 5. Effects of vitamin D derivatives on the growth of nitrosomethylurea (NMU)-induced rat mammary tumours and serum calcium concentration. Tumour-bearing rats were treated with compounds p.o. daily for 28 days at the following doses (per kilogram body weight): 0.5 µg $1,25(OH)_2D_3$, 0.1 µg KH1060, 0.5 µg KH1049, 0.1 µg MC1301, 1.0 µg CB966, 0.5 µg EB1089. *Positive values* indicate tumour progression; *negative values* show that tumours have regressed. Serum calcium concentrations at the end of the treatment period are shown in the *lower panel.*
*Significantly different from control, $p < 0.05$

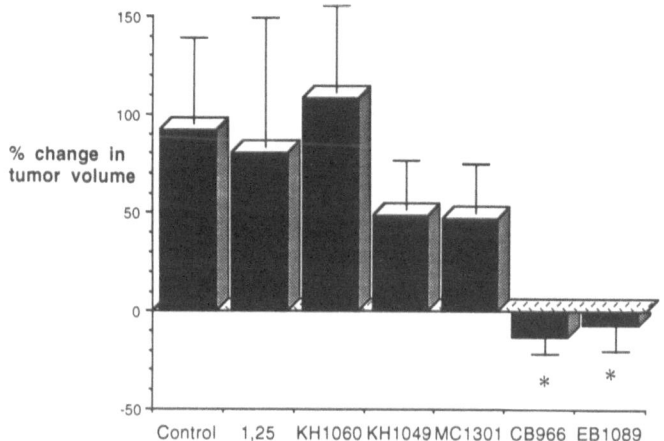

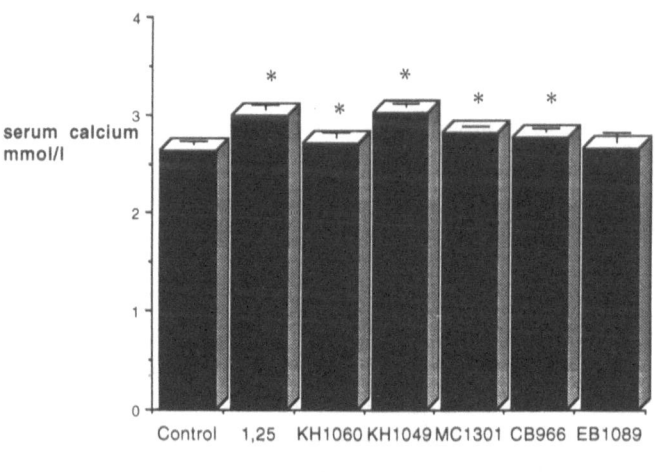

Fig. 5. Legend see p. 212

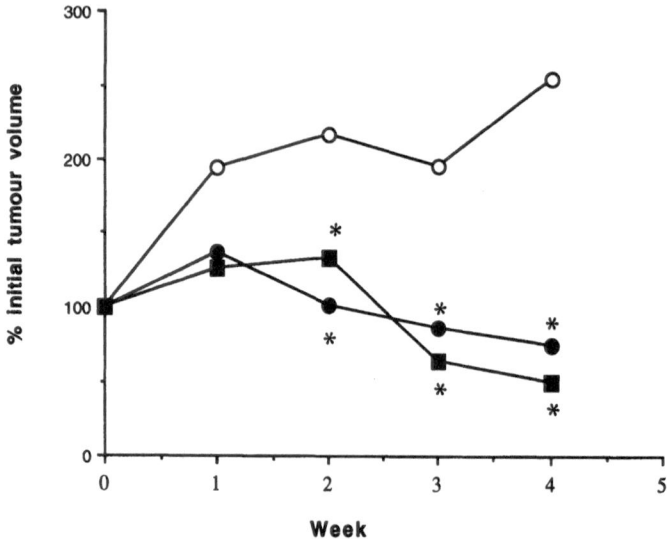

Fig. 6. Effects of EB1089 (1 µg/kg; *solid circles*) and tamoxifen (1 mg/kg; *squares*) on the growth of nitrosomethylurea (NMU)-induced rat mammary tumours. Animals were treated p.o. daily for 28 days.
*Significantly different from control (*open circles*; $p < 0.05$)

antitumour effects and decreased calcium mobilization relative to $1,25(OH)_2D_3$.

Effects of EB1089 were consistent and dose dependent. At a higher dose (1 µg/kg body weight), EB1089 produces striking tumour regression similar to that seen with the anti-oestrogen tamoxifen (Fig. 6).

With a dose of 2.5 µg/kg tumours shrank to less than half their initial volume; however, this dose caused significant hypercalcaemia (Fig. 7). In an attempt to limit the rise in serum calcium produced by this high dose of EB1089, rats were treated with EB1089 together with the bisphosphonate APD. APD alone had no effect on tumour growth, the per cent change in tumour volume was not significantly different from that in the control group. Combined treatment with EB1089 and APD led to a striking regression of tumour volume, but there was no appreciable diminution of the ability of the vitamin D analogue to raise serum calcium concentration (Fig. 7). From this we conclude that the effects of EB1089 on calcium mobilization, while considerably reduced relative

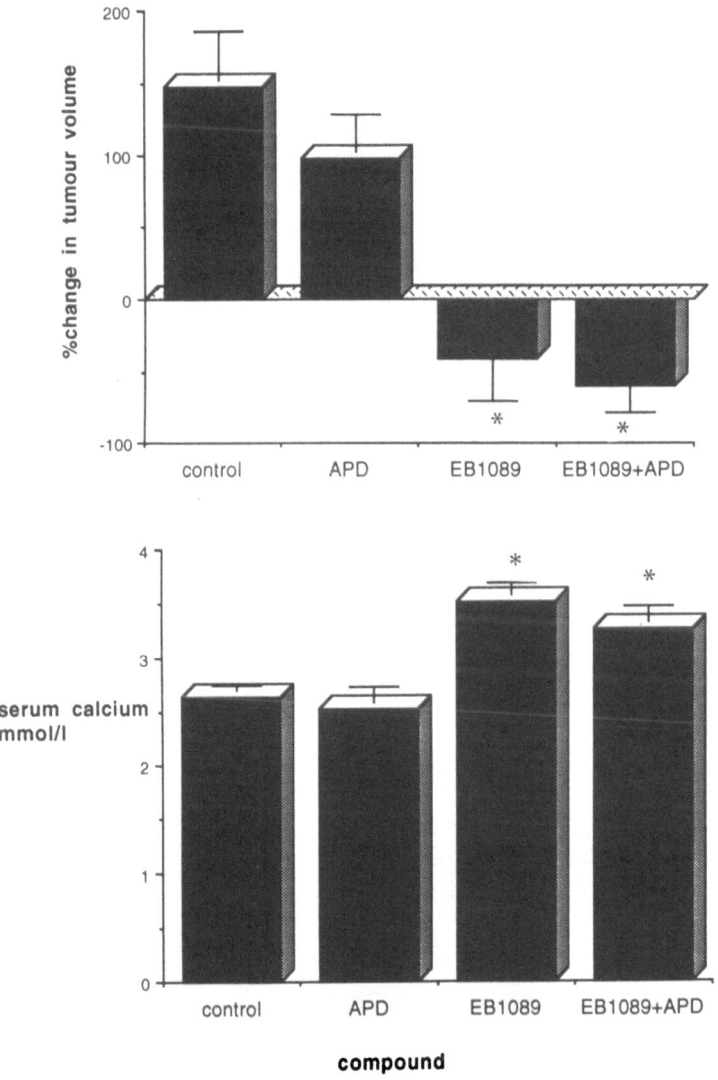

Fig. 7. Interactions of EB1089 (2.5 µg/kg) and the bisphosphonate APD (0.4 mg/kg) on growth of nitrosomethylurea (NMU)-induced rat mammary tumours (*top panel*) and serum calcium concentration (*lower panel*). *Significantly different from control, $p < 0.05$

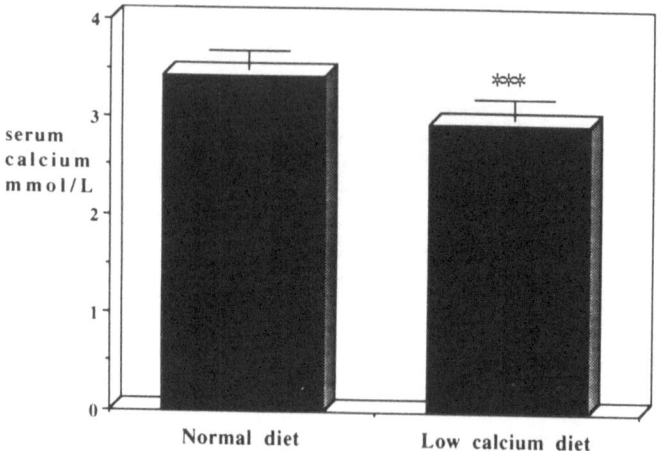

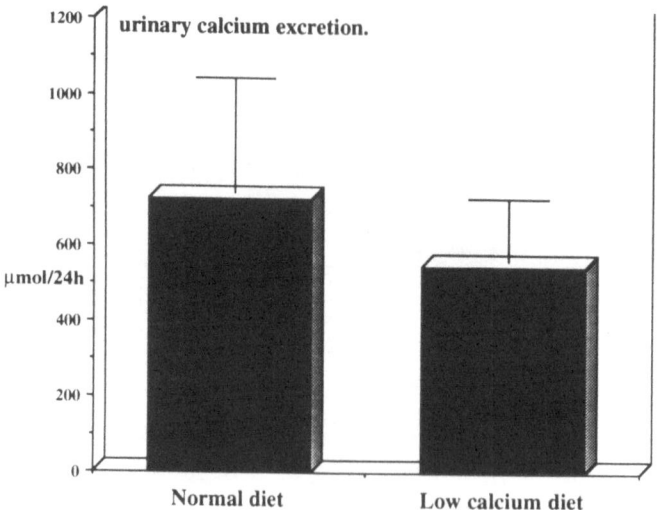

Fig. 8. Effects of normal (1 %) or low (0.1 %) calcium-containing diet on serum calcium concentration (*upper panel*) and urinary calcium excretion (*lower panel*) in normal female rats treated with EB 1089 (5 µg/kg given thrice weekly for 3 weeks). ***Significantly different from rats receiving 1 % calcium diet, $p < 0.005$

to those of 1,25(OH)₂D₃, are primarily mediated via increased calcium absorption. This was confirmed by evaluating serum calcium concentrations in normal rats treated with EB1089 maintained on a normal or low (0.1%) calcium diet. Mean serum calcium concentration in animals receiving the low calcium diet was significantly lower than in the group maintained on normal 1% calcium-containing diet (Fig. 8).

11.7 Mechanism of Action

To probe the mechanisms involved in the antiproliferative actions of these compounds, the effects of EB1089 on oestrogen response pathways have been assessed. 17β-Estradiol stimulates the proliferation of MCF-7 cells by a receptor-mediated pathway. We have evaluated ef-

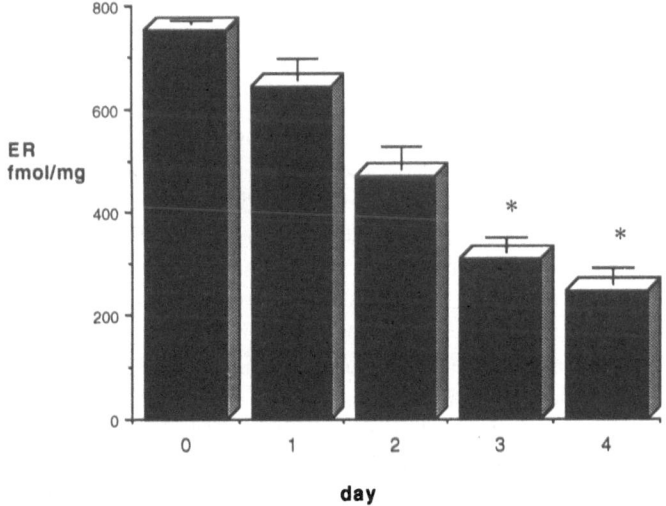

Fig. 9. Regulation of oestrogen receptor (*ER*) expression by EB1089 in MCF-7 cells. Cells grown in phenol red-free DMEM (Dulbecco's modification of Eagles minimal essential medium) medium supplemented with 5 % charcoal stripped fetal calf serum were treated for 1–4 days with $10^{-8} M$ EB1089 or ethanol vehicle as control. Cell cytosols were prepared and ER content was quantitated by the Abbott ER-EIA method. Results are expressed as mean ER concentration (fmol/mg cytosol protein) ± SEM. *Significantly different from control, $p < 0.05$

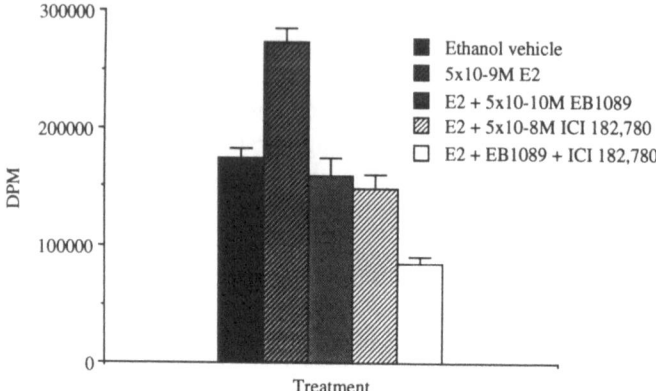

Fig. 10. Interactions of EB1089 and ICI 182,780 on 17β-estradiol (E_2) stimulated growth of MCF-7 cells. Cells were grown in phenol red-free RPMI 1640
medium supplemented with 2.5 % charcoal-stripped serum for 4 days in the
presence or absence of the indicated compounds

fects of EB1089 on the expression of oestrogen receptor (ER) in these
cells and Fig. 9 shows that EB1089 produces a dose-dependent decrease
in abundance of ER protein such that at 4 days of treatment a greater
than threefold reduction of ER levels is observed. The level of ER
transcripts is also diminished in EB1089-treated cultures and the time
course of change at the level of transcription is more rapid (data not
shown).

A consequence of the downregulation of ER by EB1089 is a diminution of the functional responsiveness of MCF-7 breast cancer cells to the
mitogenic actions of estradiol. Treatment of MCF-7 cell cultures with
combinations of estradiol and EB10989 ranging from $5 \times 10^{-11}M$ to
$5 \times 10^{-9}M$ revealed the ability of EB1089 to suppress the mitogenic
effects of estradiol in a dose-dependent manner (James et al. 1994). The
interaction of EB1089 and the pure anti-oestrogen ICI 182,780 (Wakeling et al. 1991) on the estradiol-stimulated growth of MCF-7 cells was
investigated. Treatment of cell cultures with $5 \times 10^{-10}M$ EB1089 in
combination with ICI 182,780 ($5 \times 10^{-8}M$) and in the presence of
$5 \times 10^{-9}M$ 17β-estradiol produced an augmented inhibition of proliferation compared to the actions of either compound alone (Fig. 10).

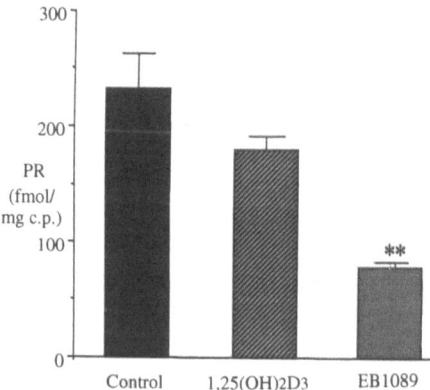

Fig. 11. Analysis of progesterone receptor protein (*PR*) in MCF-7 cells treated with EB1089. MCF-7 cells were treated with EB1089 or $1,25(OH)_2D_3$ (both at $10^{-8}M$) for 4 days. Cell cytosols were prepared and PR content was quantitated by the Abbott PR-EIA method. Results are expressed as mean PR concentration (fmol/mg) ± SEM. **Significantly different from control, $p < 0.01$

Further evidence that the vitamin D analogue limits functional responsiveness to estradiol has been obtained by assessing expression of oestrogen-responsive proteins.The progesterone receptor is an ~200-kDa protein comprising two dissimilar subunits. Its synthesis is dependent on the action of oestrogen and its expression is regulated at the level of transcription by 17β-estradiol (Ree et al. 1989). Another estradiol-regulated protein is pS2 (Rio et al. 1987). This is an ~6.5-kDa protein expressed in oestrogen-dependent breast tumours (and also stomach mucosa cells). Its function is unknown but its expression has been shown to be transcriptionally regulated by 17β-estradiol in MCF-7 cells. (Nunez et al. 1987). Our studies have shown that there is a decrease in progesterone receptor concentrations in MCF-7 cell cultures in response to treatment with EB1089. Progesterone receptor was measured by using the Abbott immunoassay (Fig. 11). Similarly, a decrease in the level of transcripts for the oestrogen-responsive pS2 protein can be demonstrated in cell cultures treated with EB1089 for 1–4 days (Fig. 12).

The mechanism by which these vitamin D derivatives cause tumour regression is quite unknown, although striking histological changes

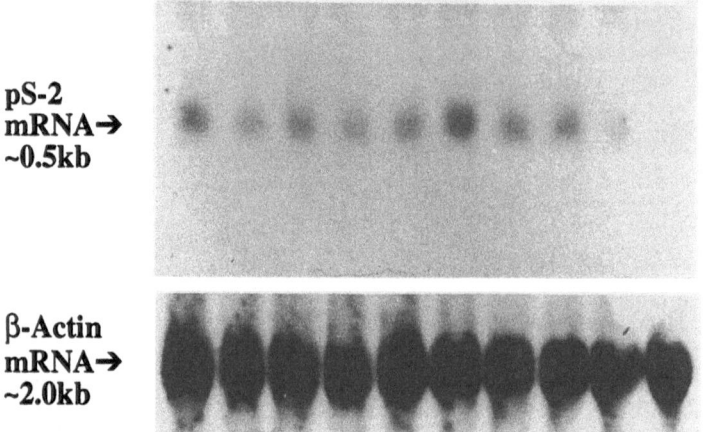

Fig. 12. Effect of EB1089 and 1,25(OH)$_2$D$_3$ on pS2 transcript levels. MCF-7 cells were cultured in phenol red-free DMEM with 5 % charcoal-stripped serum in the presence of 10^{-8}M EB1089, 10^{-8}M 1,25(OH)$_2$D$_3$ or ethanol vehicle for 1–4 days (*1–4*) prior to extraction of total RNA and northern blot analysis. (*C*, vehicle)

observed in tumours from animals treated with EB1089 suggest that specific deletion of tumour cells by a process involving programmed cell death could play a part. Furthermore, in preliminary studies we have demonstrated that treatment of oestrogen-dependent breast cancer cells with vitamin D derivatives is associated with induction of TRPM-2, a gene induced in a number of cell types undergoing apoptosis. Thus induction of TRPM-2 gene has been reported during prostatic involution following chemical or surgical castration (Monpetit et al. 1986; Léger et al. 1987) and in oestrogen-dependent MCF-7 breast tumour xenografts (Kyprianou et al. 1991) and androgen-dependent Shionogi mouse mammary carcinomas (Rennie et al. 1988) undergoing apoptosis following ablation of the respective trophic hormone. Figure 13 shows a clear increase in the level of TRPM-2 transcripts in MCF-7 breast

$$1,25(OH)_2D_3 \qquad EB1089$$

C $10^{-11}10^{-10}10^{-9}\,10^{-8}$ C $10^{-11}10^{-10}10^{-9}10^{-8}$

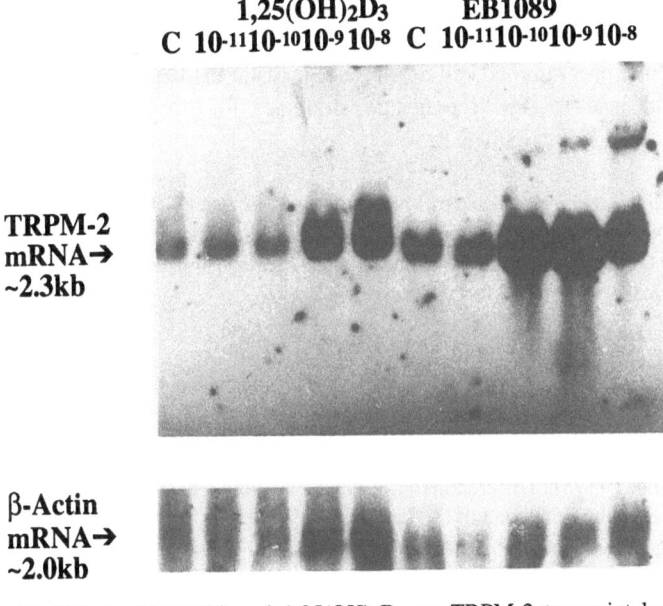

**TRPM-2
mRNA→
~2.3kb**

**β-Actin
mRNA→
~2.0kb**

Fig. 13. Effect of EB1089 and 1,25(OH)$_2$D$_3$ on TRPM-2 transcript levels. MCF-7 cells were treated with vehicle (*C*), 1,25(OH)$_2$D$_3$ or EB1089 (10^{-11} to $10^{-8}M$) for 2 days prior to extraction of total RNA and northern blot analysis

cancer cells treated for 2 days with 1,25(OH)$_2$D$_3$ and the synthetic vitamin D analogue EB1089.

Since this vitamin D analogue downregulates the expression of the oestrogen receptor in these cells and limits their responsiveness both to the mitogenic effects of 17β-estradiol and to the induction by this steroid of the progesterone receptor protein and pS2 mRNA, EB1089 is effectively producing "hormone ablation" by decreasing the oestrogen response pathways and, in this way, may promote active cell death by similar mechanisms to those reported for anti-oestrogens in vitro and estradiol withdrawal. We have also demonstrated antiproliferative effects of this and similar compounds in hormone-independent breast cancer cells (James et al. 1994). The signalling pathways through which hormone ablation induces the biochemical cascade leading to programmed cell death is unknown. Hormone-independent cancer cells retain a major portion of this cascade but display a defect in the signalling pathway such that programmed cell death is no longer activated by hormone deficiency. Identification of signalling pathways which activate the programmed cell death cascade distal to the point of this defect could lead to new therapeutic strategies for hormone-independent cancers.

Acknowledgements. This work was supported in part by the Cancer Research Campaign and the Leo Foundation. Studies with EB1089 were performed in collaboration with Dr. Lise Binderup, Leo Pharmaceutical Products, Ballerup, Denmark.

References

Abe E, Miyaura C, Sakagami H, Takeda M, Konno K, Yamazaki T, Yoshiki S, SudaT (1981) Differentiation of mouse myeloid leukemia cells induced by 1,25-dihydroxyvitamin D₃. Proc Natl Acad Sci USA 78:4990–4995

Berger U, Wilson P, McClelland R, Colston K, Haussler MR, Pike JW, Coombes, RC (1987) Immunocytochemical detection of 1,25-dihydroxyvitamin D₃ receptor in breast cancer. Cancer Res 47:6793–6795

Berger U, Wilson P, McClelland RA, Colston K, Haussler MR, Pike JW, Coombes RC et al (1988) Immunocytochemical detection of 1,25-dihydroxyvitamin D receptor in normal human tissue. J Clin Endocrinol Metab 67:607–613

Binderup L, Bramm E (1988) Effects of a novel vitamin D analogue MC903 on cell proliferation and differentiation in vitro and on calcium metabolism in vivo. Biochem Pharmacol 37:887–895

Binderup L, Latini S, Binderup E, Bretting C, Calverley M, Hansen K (1991) 20 Epi-vitamin D₃ analogues: a novel class of potent regulators of cell growth and immune responses. Biochem Pharmacol 42:1569–1575

Bower M, Colston KW, Stein RC, Hedley A, Gazet JC, Ford HT, Coombes RC (1991) Topical calcipotriol treatment in advanced breast cancer. Lancet 337:701–702

Colston KW, Colston MJ, Feldman D (1981) 1,25 dihydroxyvitamin D3 and malignant melanoma: the presence of receptors and inhibition of cell growth in culture. Endocrinology 108:1083–1086

Colston KW, Berger U, Coombes RC (1989) Possible role for vitamin D in controlling breast cancer cell proliferation. Lancet 1:185–191

Colston KW, Chander SK, Mackay AG, Coombes RC (1992) Effects of synthetic vitamin D analogues in breast cancer cell proliferation in vivo and in vitro. Biochem Pharmacol 44:673–702

Cunningham D, Gilchrist NL, Cowan RA, Soukup K (1985) Alfacalcidol as a modulator of growth of lowgrade non-Hodgkin's lymphomas. Br Med J 291:1153–1155

Dunnigan MG (1992) Vitamin D status in Asian subjects. In: Smith R (ed) Proceedings of a closed workshop in vitamin D. Practice Communications, Eastbourne UK

Eisman JA, Barkla DH, Tutton PJ (1987) Suppression of in vivo growth of human cancer solid tumor xenografts by 1,25 dihydroxyvitamin D₃. Cancer Res 47:21–25

Evans RM (1988) The steroid and thyroid hormone receptor super family. Science 240:889–895

Finch PJ, Ang L, Colston KW, Maxwell JD (1992) Blunted seasonal variation in serum 25 hydroxyvitamin D and increased risk of osteomalacia in vegetarian Asians. Clin Nutr 46:509–513

Garland CF, Garland FC (1980) Do sunlight and vitamin D reduce the likelihood of colon cancer? Int J Epidemiol 9:227–231

Garland CF, Comstock GW, Garland FC, Helsing KJ, Shaw EK, Gorham ED (1989) Serum 25-hydroxyvitamin D and colon cancer: eight year prospective study. Lancet 2:1176–1178

Garland FC, Garland CF, Gorham MPH, Young BA (1990) Geographic variation in breast cancer mortality in the United States: a hypothesis involving exposure to solar radiation. Prev Med 19:616–622

Honma Y, Hozumi M, Abe E, Konno K, Fukushima M, Hata S, Nishil Y, De-Luca HF, Suda T (1983) 1,25-dihydroxyvitamin D₃ and 1α hydroxyvitamin D₃ prolong survival time of mice inoculated with myeloid leukemia cells. Proc Natl Acad Sci USA 80:201–206

James SY, Mackay AG, Binderup L, Colston KW (1994) Effects of a new synthetic analogue, EB1089, on the oestrogen responsive growth of human breast cancer cells. J Endocrinol 141:555–563

Kelsey SM, Newland AC, Cunningham J, Makin HLJ, Coldwell RD, Mills MJ, Grant IR (1992) Sustained haematological response to high-dose oral alfacalcidol in patients with myelodysplastic syndromes. Lancet 340:316–317

Koeffler HP, Hirji K, Itri L, Southern California leukemia group (1985) 1,25 dihydroxyvitamin D₃ in vitro and in vivo effects on human preleukemic and leukemic cells. Cancer Treat Rep 69:1399–1407

Kragballe K (1989) Treatment of psoriasis by the topical application of the novel cholecalciferol analogue calcipotriol (MC903). Arch Dermatol 125:1647–1652

Kyprianou N, English HF, Davidson NE, Isaacs JT (1991) Programmed cell death during regression of the MCF-7 human breast cancer following estrogen ablation. Cancer Res 51:162–166

Léger JG, Monpetit ML, Tenniswood MP (1987) Characterization and cloning of androgen-repressed mRNAs from rat ventral prostate. Biochem Biophys Res Commun 147:196–203

Mathiasen IS, Colston KW, Binderup L (1993) EB1089, a novel vitamin D analogue, has strong antiproliferative and differentiation inducing effects on cancer cells. J Steroid Biochem Mol Med 46:365–371

McDonnell DP, Mangelsdorf DJ, Pike JW, Haussler MR, O'Malley BW (1987) Molecular cloning of complementary DNA encoding the avian receptor for vitamin D. Science 235:1214–1217

Monpetit ML, Lawless KR, Tenniswood M (1986) Androgen repressed messages in the rat ventral prostate. Prostate 8:25–36

Nunez AM, Jakolew S, Briand JP, Gaire M, Krust A, Rio M, Chambon P (1987) Characterization of the estrogen-induced pS2 protein secreted by the human breast cancer cell line MCF-7. Endocrinology 121:1759–1765

Ree AH, Landmark BF, Eskild W, Levy FO, Lahooti H, Jahnsen T, Aakvaag A, Hansson V (1989) Autologous down-regulation of messenger ribonucleic acid and protein levels for estrogen receptors in MCF-7 cells: an inverse correlation to progesterone receptor levels. Endocrinology 124:2577–2583

Rennie PS, Bruchovsky N, Buttyan R, Benson M, Cheng H (1988) Gene expression during the early phases of regression of the androgen-dependent Shionogi mouse mammary carcinoma. Cancer Res 48:6309–6312

Rio MC, Bellocq JP, Gairard B, Rasmussen UB, Krust A, Koehl C, Calderoli H, Schiff V, Renaud R, Chambon P (1987) Specific expression of the pS2 gene in subclasses of breast cancers in comparison with expression of the estrogen and progesterone receptors and the oncogene erbB2. Proc Natl Acad Sci USA 61:9243–9247

Schwartz GG, Hulka BS (1990) Is vitamin D deficiency a risk factor for prostatic cancer? (hypothesis). Anti Cancer Res 10:1307–1311

Simard A, Vorbecky J, Vobecky J (1991) Vitamin D deficiency and cancer of the breast; an unprovocative ecological hypothesis. Can J Public Health 82:300–302

Slattery ML, Sorenson AW, Ford MH (1988) Dietary calcium intake as a mitigating factor in colon cancer. Am J Epidemiol 128:504–514

Wakeling AE, Dukes M and Bowler J (1991) A potent specific pure antiestrogen with clinical potential. Cancer Res 51:3867- 3873

12 The Role of Growth Factors and Extracellular Matrix Proteases in Active Cell Death in the Prostate

M. Tenniswood, R. S. Guenette, D. Taillefer, and M. Mooibroek

12.1 Active Cell Death in the Prostate

Tissue homeostasis in the prostate requires a delicate balance between cell proliferation and active cell death (ACD) or apoptosis (Davies and Eaton 1991). This latter process has been well characterized in the prostate, which regresses after castration or administration of antian-drogens (Léger et al. 1988; Rouleau et al. 1990; Tenniswood et al. 1992), resulting in the death of approximately 80% of the epithelial cells (DeKlerk and Coffey 1978; English et al. 1985). The predominant morphological changes seen in the regressing prostate are characteristic of ACD (Sandford et al. 1984; Walker et al. 1988; Bursch et al. 1990a; Clarke 1990; Zakeri et al. 1994). ACD can be broken down into several visibly distinct stages: the precondensation stage, during which the

Table 1. Partial list of genes up-regulated after hormone ablation

Gene	Tissue	Reference
Cathepsin D	Prostate	Sensibar et al. 1990
Cathepsin B	Prostate, breast	Guenette et al. 1994d
TGF-β and	Prostate	Kyprianou and Isaacs 1988,
TGF-β receptor		1989
RVP-1	Prostate	Briehl and Miesfeld 1991
TRPM-2	Prostate ,breast	Rouleau et al. 1990;
		Guenette et al. 1994a
Tenascin	Prostate	Vollmer et al. 1994
Plasminogen activator	Prostate, breast	Freeman et al. 1990
Collagenase	Prostate	Muntzing 1981
Glutathione-S-transferase,	Prostate	Chang et al. 1987
Yb subunit		
Poly(ADP)-ribose	Prostate, breast	Guenette et al. 1994a,b
polymerase		
Hsp 27	Prostate	Guenette et al. 1994b
Hsp 70	Prostate	Buttyan et al. 1988
myc	Prostate	Buttyan et al. 1988
fos	Prostate, breast	Buttyan et al. 1988
RNase	Prostate	Engel et al. 1980

genes that are necessary for ACD are induced or recruited from other functions in the gland; the condensation stage, during which the interactions between the dying cell and its neighbors are lost as the nuclear volume and cytoplasmic volume is decreased; the fragmentation phase, during which the apoptotic cell is fragmented into several apoptotic bodies; and finally phagocytosis and degradation, during which the apoptotic bodies are phagocytosed by the neighboring cells and degraded by the lysosomal enzymes either activated in the host cell or in the apoptotic body (Bursch et al. 1990a,b). Accumulating evidence suggests that ACD is not a single phenomenon, but a series of morphologically and biochemically related processes (Gullino et al. 1972; Schweichel and Merker 1973; Ying et al. 1980; Clarke 1990; Zakeri et al. 1995). Cell death of lymphocytes and other cells of reticuloendothelial origin is dominated by changes in nuclear morphology (Cohen 1993), while ACD of secretory epithelial cells involves profound cytoplasmic changes and alterations in the cell–cell and cell–substratum interactions typical of highly organized tissues (Clarke 1990; Zakeri et al. 1995). These latter changes are often, but not always, associated with alterations in nuclear morphology and with endonuclease activation and the appearance of the characteristic DNA nucleosomal ladder (Kyprianou et al. 1988; Kyprianou and Isaacs 1988; English et al. 1989) and with lysosomal activation resulting in autophagic lysis (Bruchovsky et al. 1975; Dunn 1994). This is in marked contrast to the destruction of apoptotic lymphocytes, which are degraded following lysosomal activation in the engulfing macrophages (Walker et al. 1988; Cohen et al. 1992).

Involution of the prostate results in the elimination of most of the secretory epithelial cells of the glands and requires RNA and protein synthesis (Bruchovsky et al. 1975; Lee 1981; Tenniswood et al. 1992). R_0t analysis has suggested that between 20 and 30 mRNA species are upregulated in the regressing prostate (Montpetit et al. 1986). Some of these genes, including TRPM-2/clusterin, tissue transglutaminase, and Hsp27, are induced de novo (Léger et al. 1987; Rouleau et al. 1990; Guenette et al. 1994a,b) while others such as poly(ADP-ribose) polymerase appear to be recruited from other functions in the normal tissue. Table 1 provides a partial list of genes that are upregulated in hormone-dependent tissues during tissue regression. A few of the genes induced after hormone ablation are genes that trigger ACD. However, most others ensure that the process is completed appropriately, without leakage of the intracellular components into the extracellular environment, thus preventing activation of complement or eliciting an immune response (Kerr et al. 1972; Wyllie 1987; Walker et al. 1988; Bursch et al. 1990a). At the present time it has not been firmly established whether any of the genes listed in Table 1 can serve as specific physiological triggers of ACD. It is clear, however, that overexpression of most of the genes listed does not trigger ACD, suggesting that they are involved in the later processes of ACD.

12.2 Cell–Cell Interactions in Active Cell Death

Several lines of evidence suggest that the normal function of prostatic epithelial cells is dependent on a complex interplay between the epithelium, the extracellular matrix, and the underlying mesenchyme or stroma. Cunha's seminal work in this field using Tfm mouse model has demonstrated that the differentiation of the prostatic epithelium is critically dependent on interactions with the stroma (Thompson et al. 1986; Cunha et al. 1987). These elegant reconstruction experiments, using the embryonic mesenchyme from normal and testicular feminized mice (Tfm) combined with normal and Tfm epithelium, conclusively demonstrate that the main determinant influencing the production of a normal functional epithelium is the origin of the neighboring mesenchyme. The mesenchyme probably influences epithelial function by two different, but interacting, receptor-mediated systems: growth factors derived from the stromal compartment and components of the extracellular matrix. There is considerable evidence indicating that the growth and proliferation of epithelial cells in the prostate is influenced by factors such as epidermal growth factor (EGF), transforming growth factor-α (TGF-α), transforming growth factor-β (TGF-β), nerve growth factor (NGF) and members of the insulin-like growth factor (IGF) and fibroblast growth factor (FGF) families (Mori et al. 1990; Thompson 1990; Djakiew et al. 1991; Gleave et al. 1991; Cohen et al. 1991; Hofer et al. 1991; Wilding 1991; Story 1991; Matuo et al. 1992; Chung et al. 1992). Recent evidence has suggested that the interaction of some of these growth factors with their cognate receptors, in particular FGF-7, requires both the specific epithelial receptor and defined components of the extracellular matrix, particularly heparan sulfate (McKeehan et al. 1984; Yan et al. 1992; Kan et al. 1993).

These data have led us to suggest that in secretory epithelial cells, both the induction and completion of ACD are likely to involve the disruption of growth factor-mediated cell–cell communications and destruction of the basement membrane (Tenniswood 1986; Tenniswood et al. 1990). These data also suggest that one or more growth factors (other than testosterone) synthesized by the stroma must be responsible for the survival of the secretory epithelium (Tenniswood 1986; Tenniswood et al. 1990). Over the last 5 years one of our aims has been to identify the factor(s) responsible for such signals and to clone the genes coding for these factors.

12.3 Cloning of Genes Implicated in Active Cell Death

A number of different strategies have been used to identify genes that are expressed during regression of the rat ventral prostate after castration. The first gene to be cloned and characterized in this context, TRPM-2 (testosterone repressed prostate message), was cloned by differential hybridization (Montpetit et al. 1986; Léger et al. 1987). It has subsequently been shown that TRPM-2, or clusterin, is expressed in a wide variety of tissues undergoing ACD (Tenniswood et al. 1992). Although the expression of the gene is not confined to dying cells in other systems (particularly the testis and liver) (Griswold et al. 1986; Cheng et al. 1988; de Silva et al. 1990; Sylvester et al. 1991; Zakeri et al. 1992; Bursch et al. 1995) it clearly plays an integral part in the death of secretory epithelial cells in the prostate and mammary gland. While still not firmly established, clusterin is probably involved in controlling cholesterol efflux from the membranes of the dying cells and facilitating the transfer of cholesterol of apoproteins A-I/high-density lipoprotein (ApoA-I/HDL) (M.R. Wilson et al. 1995).

Since the induction of ACD in individual cells in the rat ventral prostate is asynchronous, the use of differential hybridization has been limited to the identification of the more abundant sequences that are induced during prostate regression. To identify other, less abundant sequences, we have used a screening strategy which we originally developed to cross screen recombinant libraries cloned into different λ-based vectors (Wong et al. 1993a). We have used this lateral cross screening to clone a number of sequences that are expressed in the ventral prostate after castration and in the lactating mammary gland after weaning. Two regression selected genes (RSG) that have been identified using this methodology are of interest because they are potentially involved in modulating the interactions of the epithelial cells with the basement membrane and/or stromal compartment. The first of these genes, RSG-2 codes for cathepsin B, while RSG-8 shows substantial homology to IGF binding protein 5 (IGFBP-5). The potential role of each of these genes in ACD is described below in the context of what is known about the involvement of extracellular matrix and growth factors in the control of ACD in the prostate.

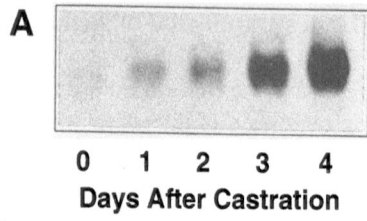

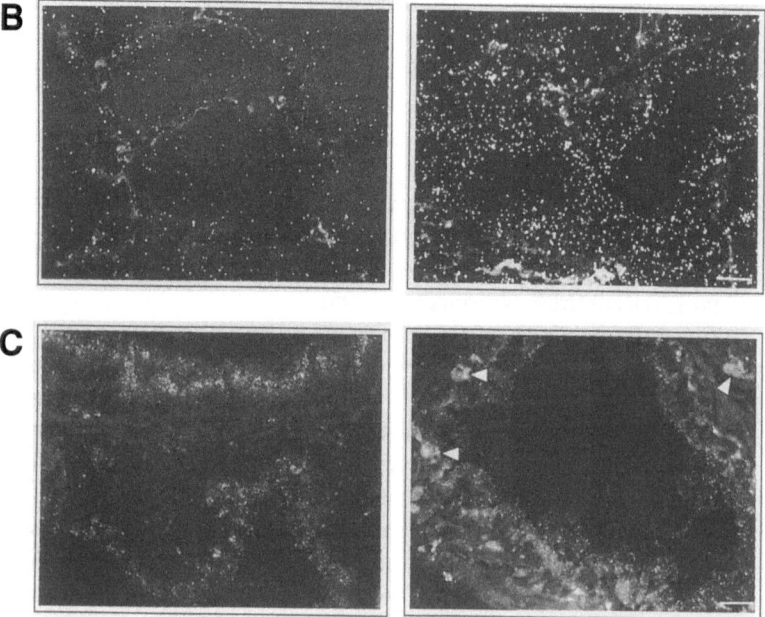

Fig. 1A–C. Expression of cathepsin B during active cell death in the prostate. **A** Northern analysis of RNA isolated from the prostate on different days after castration. **B** In situ hybridization of cathepsin B mRNA in normal rat ventral prostate prior to castration and 4 days post castration. **C** Immunofluorescent localization of cathepsin B in normal rat ventral prostate prior to castration and 4 days after castration. *Arrows* show the position of the apoptotic bodies identified under phase contrast. *Magnification bar,* 50 μm

12.4 Extracellular Matrix and Secreted Proteases in Active Cell Death

Sequence analysis of RSG-2 demonstrates that the mRNA encodes the rat homolog of cathepsin B (Guenette et al. 1994d), a thiol protease that degrades components of the basement membrane, including collagen, fibronectin, and proteoglycans (Sloane and Honn 1984). Northern analysis using RNA extracted from the rat ventral prostate at different times after castration shows that RSG-2 hybridizes to a 2-kb mRNA which is expressed at low levels in the normal prostate but is dramatically induced after androgen ablation, reaching maximal levels 4 days after castration (Fig. 1A). In situ hybridization demonstrates that cathepsin B mRNA levels in the normal rat prostate are barely above background, but are induced significantly in the dying luminal epithelial cells after castration (Fig. 1B). The protein is seen as faint but distinct punctate staining over the lysosomal compartment of the luminal epithelial cells in the normal prostate. The level of cathepsin B increases significantly in the prostate 2 and 4 days after castration and is localized predominantly in the apoptotic cells of the gland (Fig. 1C). The localization of cathepsin B to the lysosomes of the dying cells further suggests that lysosomes activated within the dying cells play an active role in the degradation of the apoptotic bodies (Dunn 1994).

The secretory cells of the prostate interact with the basement membrane located between the stroma and epithelium. This extracellular matrix contains a number of components including fibronectin, collagen, laminin, vitronectin, heparan sulfate, and chondroitin sulfate (Kofoed et al. 1990; Paulsson 1992). These components are known to interact with their cognate receptors, many of which are members of the integrin superfamily (Damsky and Werb 1992; Hynes 1992; Juliano and Haskil 1993). These receptors are localized on the basal surface of epithelial cells to ensure that they have the structural underpinnings needed for polarization, vectorial transport, and secretion (Getzenberg et al. 1990; Ginsberg et al. 1990; Pienta et al. 1991b; Petersen et al. 1992). While the composition of the basement membrane is not static, drastic changes in the composition may influence gene expression, cellular function, and motility (Lee et al. 1984; Pienta et al. 1991a; Streuli et al. 1993), and for this reason the proteolysis of the components of the basement membrane must be tightly regulated. The extracellular

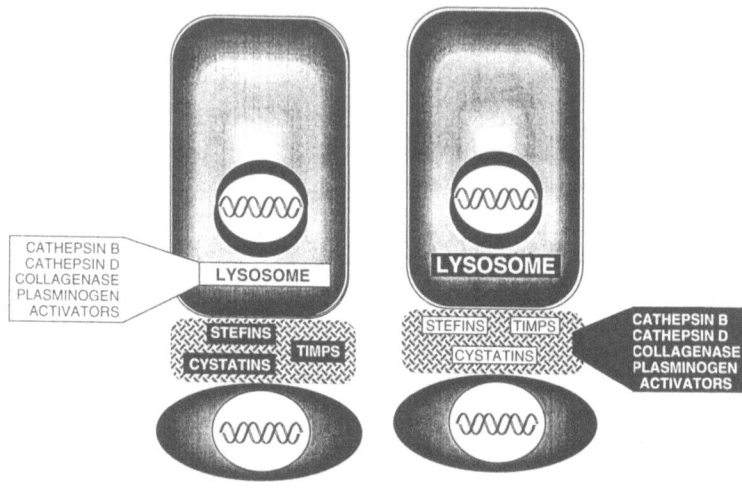

HOMEOSTASIS   ACTIVE CELL DEATH

Fig. 2. Schematic showing activation of lysosomal proteases, and inactivation of protease inhibitors during active cell death

matrix contains a number of specific protease inhibitors, including members of the tissue inhibitors of metalloprotease (TIMP) (Lokeshwar et al. 1993), cystatin, and stefin families (Lenarcic et al. 1991; Lah et al. 1992). The balance between the relative level of proteases and their specific inhibitors dictates the degree of protease activity and essentially ensures that uncontrolled proteolysis of the components of the basement membrane does not occur. Since the degradation of the basement membrane is a prerequisite for the cytoplasmic condensation of the dying cell, induction and activation of secreted proteases is a necessary part of ACD of secretory epithelial cells. The expression of cathepsin B during ACD is coincident with the expression of several other extracellular proteases, including matrix metalloproteinases 2 and 9, collagenase, tissue type and urokinase type plasminogen activators, and cathepsin D (Muntzing et al. 1979; Sensibar et al. 1990; Freeman et al. 1990; Lokeshwar et al. 1993; Wilson et al. 1994). These proteases form a proteolytic cascade that initially degrades the proteases inhibitors, overriding their effects, and subsequently degrades many components of the basement membrane, substantially altering signal transduction in the dying cell immediately prior to cellular condensation (Damsky and Werb 1992; Juliano and Haskil 1993) (Fig. 2). For example, cathepsin D actively degrades cystatin C, the major inhibitor of cathepsin B, resulting in the enhancement of the enzymatic activity of cathepsin B (Lenarcic et al. 1991) which itself activates urokinase-type plasminogen activator (Kobayashi et al. 1991) as well as degrading collagen type IV, laminin and fibronectin (Buck et al. 1992). Urokinase-type plasminogen activator in turn activates plasmin which degrades other components of the extracellular matrix. Coupled with changes in the intracellular pH and free calcium levels, which appears to activate chromatin condensation (Kyprianou et al. 1988; Arends et al. 1990; Schwartz et al. 1991; Barry and Eastman 1992), activation of tissue transglutaminase, which cross links the cytoskeletal components of the cell (Fesus et al. 1991; Guenette et al. 1994b), expression of TRPM-2/clusterin, which facilitates membrane remodeling (Wilson et al. 1994), and the exposure of vitronectin on the cell surface which stimulates phagocytosis by macrophage (Savill et al. 1990, 1993), these changes in extracellular protease activity appear to be central to mechanistic completion of ACD (Bursch et al. 1990a).

12.5 Insulin-Like Growth Factors and Active Cell Death

Sequence analysis of the second RSG we have characterized, RSG-8, demonstrates that it codes IGFBP-5 (Guenette et al. 1994c). The cDNA hybridizes to an mRNA of approximately 6.0 kb which is expressed at very low levels in the normal prostate but shows a dramatic induction after androgen ablation (Fig. 3A). Expression of IGFBP-5 mRNA is induced in the secretory epithelial cells after castration, reaching a maximum level on day 4 (Fig. 3B). The prostatic stroma is known to secrete IGF-I (Cohen et al. 1991; Iwamura et al. 1993), and the epithelial cells respond to IGF-I (and to a lesser extent IGF-II and insulin) through the high-affinity interaction of these growth factors with the type I IGF receptor (Cohen et al. 1991; Iwamura et al. 1993). This interaction is probably modulated by IGFBP-2 which presents IGF-I to the receptor and which appears to be constitutively expressed in the normal rat ventral prostate (Cohen et al. 1991; Guenette et al. 1994c).

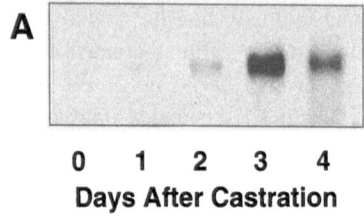

Days After Castration

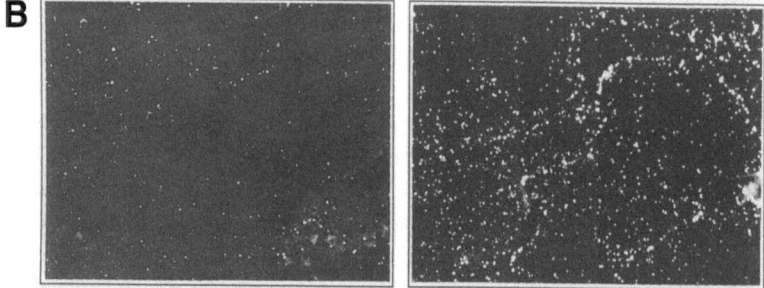

Fig. 3A,B. Expression of insulin-like growth factor binding protein (IGFBP-5) during ACD in the prostate. **A** Northern analysis of RNA isolated from the prostate on different days after castration. **B** In situ hybridization of IGFBP-5 mRNA in normal rat ventral prostate prior to castration and 4 days after castration

During early development IGFBP-2 and IGFBP-5 are coexpressed in tissues of ectodermal origin (Green et al. 1994); however, in the adult prostate IGFBP-2 is expressed while the IGFBP-5 expression is repressed. During ACD, on the other hand, the expression of IGFBP-2 mRNA is probably downregulated while IGFBP-5 mRNA expression is induced (Guenette et al. 1994c). IGFBP-2 has an RGD sequence located in the carboxy terminus and is thought to associate with integrins after secretion from the basal surface of the epithelial cells (Shimasaki and Ling 1991). In this way IGFBP-2 may serve as a storage pool for IGF-I, potentiating the effect of IGF-I by facilitating interactions with the type I IGF receptor in a manner that is similar to the proposed role of IGFBP-2 in small cell lung carcinoma (Reeve et al. 1993). During ACD the induction of IGFBP-5 expression, coupled with loss of IGFBP-2

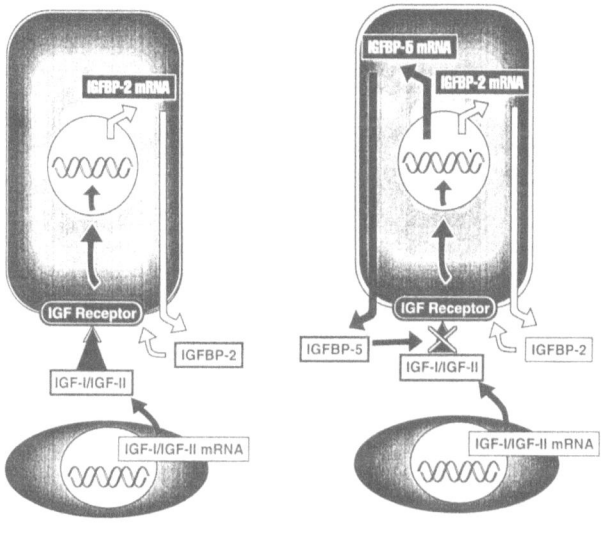

HOMEOSTASIS   **ACTIVE CELL DEATH**

Fig. 4. Schematic showing role of insulin-like growth factor binding protein (*IGFBP-2*) and (*IGFBP-5*) in the modulation and attenuation of *IGF* signaling in the rat ventral prostate during the initiation of active cell death (ACD). In the normal rat ventral prostate IGFBP-2 presents IGF-I to the IGF receptor and modulates IGF-I action. After androgen ablation IGFBP-5 mRNA is induced, and IGFBP-5 binds to IGF-I, sequestering the growth factor in the extracellular matrix and attenuating the interaction between IGF-I and the receptor, inducing ACD

expression, attenuates the cellular response to IGF-I. High-affinity binding of IGF-I to IGFBP-5, which associates with several components of the extracellular matrix including types III and IV collagen, laminin, and fibronectin (Clemmons 1993; Jones et al. 1993), sequesters IGF-I away from the receptor and appears to interfere with the normal homeostatic intracellular signaling downstream of the receptor. Close examination of the relationship between IGFBP-5, IGF-I, and ACD suggest that the expression of IGFBP-5 is an early event in ACD and may serve to trigger the process in prostate secretory epithelial cells (Fig. 4). The accumulated data suggest that in the normal prostate the expression of IGFBP-5 is directly repressed by androgens and that

IGFBP-2 serves to modulate the interaction of IGF-I and the IGF-I receptor. After androgen ablation the expression of IGFBP-5 is rapidly induced. After secretion into the extracellular matrix, IGFBP-5 competes with IGFBP-2 for IGF-I binding and through its interaction with the extracellular matrix (Jones et al. 1993) (rather than the integrins) essentially sequesters IGF-I away from the receptor. This effectively attenuates the cellular response to IGF-I which appears to be necessary for survival. The changes in second messenger signaling downstream of the receptor induces a cascade of other genes that are required later in the process, including cathepsin B and TRPM-2, that are required for ACD. It is of interest that IGFBP-5 can be proteolytically degraded by several of the extracellular proteases that are induced during ACD, including cathepsin D. This effectively limits the duration of IGFBP-5 action and ensures that stromal–epithelial interactions between neighboring cells are not inadveretently interrupted by diffusion of the binding protein.

12.6 Control of Gene Expression During Active Cell Death

The coordination of gene expression during ACD must be complex since the genes are induced in a temporally and anatomically restricted manner in the dying cells. Computer-aided analysis of the 5' promoter regions of the murine cathepsin B (Qian et al. 1991), human and rat TRPM-2/clusterin (Wong et al. 1993b, 1994), and rat IGFBP-2 (Boisclair et al. 1993) and IGFBP-5 (Zhu et al. 1993) genes does not demonstrate any obvious common DNA binding motifs. Unlike the TRPM-2/clusterin and IGFBP-5 genes, which are repressed in the normal cell (Léger et al. 1987; Guenette et al. 1994c), the cathepsin B and IGFBP-2 genes lack the TATA and CAAT motifs and appears to be constitutively expressed at low levels. In the case of cathepsin B, the levels of the mRNA appear to be increased and the protein is recruited during ACD, presumably as a result of additional DNA–protein interactions (Qian et al. 1991). Both clusterin/TRPM-2 and IGFBP-5 contain several common motifs, including AP-1 and AP-2 motifs; however, the spatial organization of these motifs in the respective promoters is quite different and could reflect differences in the timing of the induction of these genes during ACD. Since the induction of these two genes may be controlled independently, as we have suggested, this is probably reasonable. None of the genes so far implicated in the activation of ACD in the prostate have clearly identifiable androgen regulatory elements, such as have been identified in the prostate steroid binding protein gene (Claessens et al. 1989), raising the possibility that all of these genes are downstream of another criticial "trigger" gene that has yet to be identified. While the promoter regions of the cathepsin B and TRPM-2 are quite different, it is of interest that both genes are localized to the same region of chromosome 8–8p22 in the case of cathepsin B (Fong et al. 1986) and 8p21 in the case of TRPM-2/clusterin (Fink et al. 1993; Wong et al. 1994), whereas IGFBP-2 is localized to chromosome 2q33-34 and IGFBP-5 is localized on chromosome 5 (Shimasaki and Ling 1991). This raises the possibility that the induction of the genes involved in the later stages of ACD may be coordinated at higher levels of chromatin structure, rather than through specific *cis*-acting domains in the respective promoter regions.

12.7 Active Cell Death and Metastasis

A number of studies have demonstrated that the metalloproteases and cathepsins play an important functional role in metastatic disease. The level of protease activity is elevated in human breast and ovarian cancers (Woynarowska et al. 1989; Gabrijelcic et al. 1992; Lah et al. 1992) and in many metastatic cell lines, including ras-transformed NIH 3T3 cells and Lewis lung carcinoma cells, in which the level of expression has been correlated with the increased malignancy (Brodt et al. 1992; Chambers et al. 1992). In many malignant cell lines the enzyme is localized to the plasma membrane and is clearly secreted in most instances (Sloane and Honn 1984; Sloane et al. 1987, 1991). The level of the protease inhibitors, on the other hand, are frequently decreased in metastatic cell lines (Liotta and Stetler-Stevenson 1990).

These data, taken together, reinforce the observation that many of the features of ACD are recapitulated in metastasis and raise the possibility that "arrested" ACD may lead to metastatic progression. The major distinction in the formation of an apoptotic cell and a metastatic cell lies in the destruction of the DNA as a result of the activation of one or more endogenous endonucleases (Tenniswood et al. 1992). In the absence of endonuclease activation, possibly as a result of oncogenic mutation which disrupts the signaling pathways necessary for induction of ACD, the proteases induced during the precondensation phase of ACD may also serve to increase the motility of the cell that escapes from the ACD pathway. These cells will have an enhanced capacity for the degradation of the ECM, and thus may have the potential to extravasate from the tissue. Further experiments are clearly needed to test this hypothesis and to determine whether the metastatic phenotype can develop as a result of "arrested" ACD and to obtain a clearer picture of the sequence of events that occurs in individual cells as they undergo ACD.

Acknowledgements. Some of the work described in this manuscript forms parts of the Ph.D. thesis research of RSG and DT, in the Department of Biochemistry, University of Ottawa, Ottawa, Ontario, Canada. This work was funded, in part, by operating grants to MT from the Medical Research Council of Canada, and the National Cancer Institute of Canada.

References

Arends MJ, Morris RG, Wyllie AH (1990) Apoptosis. The role of the endonuclease. Am J Pathol 136:593–608

Barry MA, Eastman A (1992) Endonuclease activation during apoptosis: the role of cytosolic Ca^{2+} and pH. Biochem Biophys Res Commun 186:782–789

Boisclair YR, Brown AL, Casola S, Rechler MM (1993) Three clustered Sp1 sites are required for efficient transcription of the TATA-less promoter of the gene for insulin-like growth factor binding protein-2 from the rat. J Biol Chem 268:24892–24901

Briehl MM, Miesfeld RL (1991) Isolation and characterization of transcripts induced by androgen withdrawal and apoptotic cell death in the rat ventral prostate. Mol Endocrinol 5:1381–1388

Brodt P, Reich R, Moroz LA, Chambers AF (1992) Differences in the repertoires of basement membrane degrading enzymes in two carcinoma sublines with distinct patterns of site-selective metastasis. Biochim Biophys Acta 1139:77–83

Bruchovsky N, Lesser B, van Doorn E, Craven S (1975) Hormonal effects on cell proliferation in rat prostate. Vit Horm Res 33:61–100

Buck MR, Karustis DG, Day NA, Honn KV, Sloane BF (1992) Degradation of extracellular-matrix proteins by human cathepsin B from normal and tumor tissues. Biochem J 282:273–278

Bursch W, Kleine L, Tenniswood MP (1990a) The biochemistry of cell death by apoptosis. Biochem Cell Biol 68:1071–1074

Bursch W, Paffe S, Putz B, Barthel G, Schulte-Hermann R (1990b) Determination of the length of the histological stages of apoptosis in normal liver and in altered hepatic foci of rats. Carcinogenesis 11:847–853

Bursch W, Gleeson TG, Kleine L, Tenniswood MP (1995) Expression of TRPM-2/clusterin mRNA during growth and regression of rat liver. Arch Toxicol (in press)

Buttyan R, Zakeri Z, Lockshin R, Wolgemuth D (1988) Cascade induction of c-fos, c-myc, and heat shock 70 K transcripts during regression of the rat ventral prostate gland. Mol Endocrinol 2:650–657

Chambers AF, Colella R, Denhardt DT, Wilson SM (1992) Increased expression of cathepsins L and B and decreased activity of their inhibitors in metastatic, ras-transformed NIH 3T3 cells. Mol Carcinog 5:238–245

Chang C, Saltzman AG, Sorensen NS, Hiipakka RA, Liao S (1987) Identification of gluatathione S-transferase Yb1 mRNA as the androgen-repressed mRNA by cDNA cloning and sequence analysis. J Biol Chem 262:11901–11903

Cheng CY, Chen CL, Feng ZM, Marshall A, Bardin CW (1988) Rat clusterin isolated from primary Sertoli cell-enriched culture medium is sulfated glycoprotein-2 (SGP-2). Biochem Biophys Res Commun 155:398–404

Chung LW, Li W, Gleave ME, Hsieh JT, Wu HC, Sikes RA, Zhau HE, Bandyk MG, Logothetis CJ et al (1992) Human prostate cancer model: roles of growth factors and extracellular matrices. J Cell Biochem [Suppl] 16H:99–105

Claessens F, Celis L, Peeters B, Heyns W, Verhoeven G, Rombauts W (1989) Functional characterization of an androgen response element in the first intron of the C3(1) gene of prostatic binding protein. Biochem Biophys Res Commun 164:833–840

Clarke PG (1990) Developmental cell death: morphological diversity and multiple mechanisms. Anat Embryol 181:195–213

Clemmons DR (1993) IGF binding proteins and their functions. Mol Reprod Dev 35:368–375

Cohen JJ (1993) Apoptosis. Immunol Today 14:126–130

Cohen JJ, Duke RC, Fadok VA, Sellins KS (1992) Apoptosis and programmed cell death in immunity. Annu Rev Immunol 10:267–293

Cohen P, Peehl DM, Lamson G, Rosenfeld RG (1991) Insulin-like growth factors (IGFs), IGF receptors, and IGF-binding proteins in primary cultures of prostate epithelial cells. J Clin Endocrinol Metab 73:401–407

Cunha GR, Donjacour AA, Cooke PS, Mee S, Bigsby RM, Higgins SJ, Sugimura Y (1987) The endocrinology and developmental biology of the prostate. Endocr Rev 8:338–362

Damsky CH, Werb Z (1992) Signal transduction by integrin receptors for extracellular matrix: co-operative processing of extracellular information. Curr Opin Cell Biol 4:772–781

Davies P, Eaton CL (1991) Regulation of prostate growth. J Endocrinol 131:5–17

de Silva HV, Harmony JAK, Stuart WD, Gil CM, Robbins J (1990) Apolipoprotein J: structure and tissue distribution. Biochemistry 29:5380–5389

DeKlerk DP, Coffey DS (1978) Quantitative determination of prostatic epithelial and stromal hyperplasia by a new technique. Invest Urol 16:240–245

Djakiew D, Delsite R, Pflug B, Wrathall J, Lynch JH, Onoda M (1991) Regulation of growth by a nerve growth factor-like protein which modulates paracrine interactions between a neoplastic epithelial cell line and stromal cells of the human prostate. Cancer Res 51:3304–3310

Dunn WAJ (1994) Autophagy and related mechanisms of lysosome-mediated protein degradation. Trends Cell Biol 4:139–143

Engel G, Lee C, Grayhack JT (1980) Acid ribonuclease in rat prostate during castration-induced involution. Biol Reprod 22:827–831

English HF, Drago JR, Santen RJ (1985) Cellular response to androgen depletion and repletion in the rat ventral prostate: autoradiography and morphometric analysis. Prostate 7:41–51

English HF, Kyprianou N, Isaacs JT (1989) Relationship between DNA fragmentation and apoptosis in the programmed cell death in the rat prostate following castration. Prostate 15:233–250

Fesus L, Tarcsa E, Kedei N, Autuori F, Piacentini M (1991) Degradation of cells dying by apoptosis leads to accumulation of epsilon(gamma-glutamyl)lysine isodipeptide in culture fluid and blood. FEBS Lett 284:109–112

Fink TM, Zimmer M, Tschopp J, Etienne J, Jenne DE, Lichter P (1993) Human clusterin (CLI) maps to 8p21 in proximity to lipoprotein lipase (LPL) gene. Genomics 16:526–528

Fong D, Calhoun DH, Hsieh WT, Lee B, Wells RD (1986) Isolation of a cDNA clone for the human lysosomal proteinase cathepsin B. Proc Natl Acad Sci USA 83:2909–2913

Freeman SN, Rennie PS, Chao J, Lund LR, Andreasen PA (1990) Urokinase- and tissue-type plasminogen activators are suppressed by cortisol in the involuting prostate of castrated rats. Biochem J 269:189–193

Gabrijelcic D, Svetic B, Spaic D, Skrk J, Budihna M, Dolenc I, Popovic T, Cotic V, Turk V (1992) Cathepsins B, H and L in human breast carcinoma. Eur J Clin Chem Clin Biochem 30:69–74

Getzenberg RH, Pienta KJ, Coffey DS (1990) The tissue matrix: cell dynamics and hormone action. Endocr Rev 11:399–417

Ginsberg MH, Loftus JC, D'Souza S, Plow EF (1990) Ligand binding to integrins: common and ligand specific recognition mechanisms. Cell Differ Dev 32:203–214

Gleave M, Hsieh JT, Gao CA, von Eschenbach AC, Chung LW (1991) Acceleration of human prostate cancer growth in vivo by factors produced by prostate and bone fibroblasts. Cancer Res 51:3753–3761

Green BN, Jones SB, Streck RD, Wood TL, Rotwein P, Pintar JE (1994) Distinct expression patterns of insulin-like growth factor-like binding protein 2 and 5 during fetal and postnatal development. Endocrinology 134:954–962

Griswold MD, Roberts K, Bishop P (1986) Purification and characterization of a sulfated glycoprotein secreted by Sertoli cells. Biochem 25:7265–7270

Guenette RS, Corbeil HB, Leger JG, Wong K, Mezl V, Mooibroek M, Tenniswood MP (1994a) Induction of gene expression during involution of the lactating rat mammary gland. J Mol Endocrinol 12:47–60

Guenette RS, Daehlin L, Mooibroek M, Wong K, Tenniswood M (1994b) Thanatogen expression during involution of the rat ventral prostate after castration. J Androl 15:200–211

Guenette RS, Mooibroek M, Wong K, Tenniswood M (1994c) Regulation of insulin like growth factor binding protein expression during active cell death in hormone dependent tissues. Cell Growth Differ (in preparation)

Guenette RS, Mooibroek M, Wong K, Wong P, Tenniswood M (1994d) Cathepsin B, a cysteine protease implicated in metastatic progression, is also expressed during regression of the rat prostate and mammary glands. Eur J Biochem 226:311–321

Gullino PM, Grantham FH, Losonczy I, Berghoffer B (1972) Mammary tumor regression. I. Physiopathologic characteristics of hormone dependent tissue. J Natl Cancer Inst 49:1333–1348

Hofer DR, Sherwood ER, Bromberg WD, Mendelsohn J, Lee C, Kozlowski JM (1991) Autonomous growth of androgen-independent human prostatic carcinoma cells: role of transforming growth factor alpha. Cancer Res 51:2780–2785

Hynes RO (1992) Integrins: versatility, modulation, and signalling in cell adhesion. Cell 69:11–25

Iwamura M, Sluss PM, Cassmento JB, Cockett ATK (1993) Insulin-like growth factor I: action and receptor characterization in human prostate cancer cell lines. Prostate 22:243–252

Jones J, Gockerman A, Busby WH, Camacho-Hubner C, Clemmons DR (1993) Extracellular matrix contains insulin-like growth factor binding protein-5: potentiation of the effects of IGF-1. J Cell Biol 121:679–687

Juliano RL Haskil S (1993) Signal transduction from the extracellular matrix. J Cell Biol 120:577–585

Kan M, Wang F, Xu J, Crabb JW, Hou J, McKeehan WL (1993) An essential heparin-binding domain in the fibroblast growth factor receptor kinase. Science 259:1918–1921

Kerr JFR, Wyllie AH, Currie AR (1972) Apoptosis: a basic biological phenomenon with wide ranging implications in tissue kinetics. Br J Cancer 26:239–257

Kobayashi H, Schmitt M, Goreetzki L, Chucholowski N, Calvete J, Kramer M, Gunzler WA, Janicke F, Graeff H (1991) Cathepsin B efficiently activates the soluble and the tumor cell receptor bound form of the proenzyme urokinase-type plasminogen activator (pro-uPA). J Biol Chem 266:5147–5152

Kofoed JJ, Tumilasci OR, Curbelo HM, Fernandez Lemos SM, Arias NH, Houssay AB (1990) Effects of castration and androgens upon prostatic proteoglycans in rats. Prostate 16:93–102

Kyprianou N, Isaacs JT (1988) Activation of programmed cell death in the rat ventral prostate after castration. Endocrinol 122:552–562

Kyprianou N Isaacs JT (1989) Expression of transforming growth factor-beta in the rat ventral prostate during castration-induced programmed cell death. Mol Endocrinol 3:1515–1522

Kyprianou N, English HF, Isaacs JT (1988) Activation of a Ca^{2+}-Mg^{2+}-dependent endonuclease as an early event in castration-induced prostatic cell death. Prostate 13:103–117

Lah TT, Kokalj-Kunovar M, Strukelj B, Pungercar J, Barlic-Maganja D, Drobnic-Kosorok M, Kastelic L, Babnik J, Golouh R et al (1992) Stefins and lysosomal cathepsins B, L and D in human breast carcinoma. Int J Cancer 50:36–44

Lee C (1981) Physiology of castration-induced regression of the rat prostate. Proc Clin Biol Res 75A:145–159

Lee EY-H, Parry G, Bissell MJ (1984) Modulation of secreted proteins of mouse mammary epithelial cells by the collagenous substrata. J Cell Biol 98:146–155

Léger JG, Montpetit ML, Tenniswood MP (1987) Characterization and cloning of androgen-repressed mRNAs from rat ventral prostate. Biochem Biophys Res Commun 147:196–203

Léger JG, Le Guellec R, Tenniswood MP (1988) Treatment with antiandrogens induces an androgen-repressed gene in the rat ventral prostate. Prostate 13:131–142

Lenarcic B, Krasovec M, Ritonja A, Olafsson I, Turk V (1991) Inactivation of human cystatin C and kininogen by human cathepsin D. FEBS Lett 280:211–215

Liotta L, Stetler-Stevenson WG (1990) Metalloproteinases and cancer invasion. Semin Cancer Biol 1:107–115

Lokeshwar BL, Selzer MG, Block NL, Gunja-Smith Z (1993) Secretion of matrix metalloproteinases and their inhibitors (tissue inhibitors of metalloproteinases) by human prostate in explant cultures: reduced tissue inhibitor of metalloproeinase secretion by malignant tissues. Cancer Res 53:4493–4498

Matuo Y, McKeehan WL, Yan GC, Nikolaropoulos S, Adams PS, Fukabori Y, Yamanaka H, Gaudreau J (1992) Potential role of HBGF (FGF) and TGF-beta on prostate growth. Adv Exp Med Biol 324:107–114

McKeehan WL, Adams PS, Rosser MP (1984) Direct mitogenic effects of insulin, epidermal growth factor, glucocorticoid, cholera toxin, unknown pituitary factors and possibly prolactin, but not androgen, on normal rat prostate epithelial cells in serum-free, primary cell culture. Cancer Res 44:1998–2010

Montpetit ML, Lawless KR, Tenniswood M (1986) Androgen-repressed messages in the rat ventral prostate. Prostate 8:25–36

Mori H, Maki M, Oishi K, Jaye M, Igarashi K, Yoshida O, Hatanaka M (1990) Increased expression of genes for basic fibroblast growth factor and transforming growth factor type beta 2 in human benign prostatic hyperplasia. Prostate 16:71–80

Muntzing J (1981) Collagen synthesis and breakdown in the rat ventral prostate. In: Liss AR (ed) The prostatic cell: structure and function. Liss, New York, pp 137–144

Muntzing J, Liljekvist J, Murphy GP (1979) Chalones and stroma as possible growth-limiting factors in the rat ventral prostate. Invest Urol 16:399–402

Paulsson M (1992) Basement membrane proteins: structure, assembly, and cellular interactions. Crit Rev Biochem Mol Biol 27:93–127

Petersen OW, Ronnov-Jessen L, Howlett AR, Bissell MJ (1992) Interaction with basement membrane serves to rapidly distinguish growth and differentiation pattern of normal and malignant human breast epithelial cells. Proc Natl Acad Sci USA 89:9064–9068

Pienta KJ, Isaacs WB, Vindivich D, Coffey DS (1991a) The effects of basic fibroblast growth factor and suramin on cell motility and growth of rat prostate cancer cells. J Urol 145:199–202

Pienta KJ, Murphy BC, Getzenberg RH, Coffey DS (1991b) The effect of extracellular matrix on morphologic transformation in vitro. Biochem Biophys Res Commun 179:333–339

Qian F, Frankfater A, Chan SJ, Steiner DF (1991) The structure of the mouse cathepsin B gene and its putative promoter. DNA Cell Biol 10:159–168

Reeve JG, Morgan J, Schwander J, Bleehen NM (1993) Role for membrane and secreted insulin-like growth factor-binding protein-2 in the regulation of insulin-like growth factor action in lung tumors. Cancer Res 53:4680–4685

Rouleau M, Léger JG, Tenniswood MP (1990) Ductal heterogeneity of cy-
    tokeratins, gene expression and cell death in the rat ventral prostate. Mol
    Endocrinol 4:2003–2013
Sandford NL, Searle JW, Kerr JFR (1984) Successive waves of apoptosis in
    the rat prostate after repeated withdrawal of testosterone. Pathology
    16:406–410
Savill J, Dransfield I, Hogg N, Haslett C (1990) Vitronectin receptor-mediated
    phagocytosis of cells undergoing apoptosis. Nature 343:170–173
Savill J, Fadok V, Henson P, Haslett C (1993) Phagocyte recognition of cells
    undergoing apoptosis. Immunol Today 14:131–136
Schwartz MA, Ingber DE, Lawrence M, Springer TA, Lechene C (1991)
    Multiple integrins share the ability to induce elevation of intracellular pH.
    Exp Cell Res 195:533–535
Schweichel J-U, Merker HJ (1973) The morphology of various types of cell
    death in prenatal tissues. Teratology 7:253–266
Sensibar JA, Liu XX, Patai B, Alger B, Lee C (1990) Characterization of cas-
    tration-induced cell death in the rat prostate by immunohistochemical local-
    ization of cathepsin D. Prostate 16:263–276
Shimasaki S, Ling N (1991) Identification and molecular characterization of
    insulin-like growth factor binding proteins (IGFBP-1, -2, -3, -4, -5 and -6).
    Prog Growth Factor Res 3:243–266
Sloane BF Honn KV (1984) Cysteine proteinases and metastasis. Cancer
    Metastasis Rev 3:249–263
Sloane BF, Rozhin J, Hatfield JS, Crissman JD, Honn KV (1987) Plasma
    membrane-associated cysteine proteinases in human and animal tumors.
    Exp Cell Biol 55:209–224
Sloane BF, Rozhin J, Krepela E, Ziegler G, Sameni M (1991) The malignant
    phenotype and cysteine proteinases. Biomed Biochim Acta 50:549–554
Story MT (1991) Polypeptide modulators of prostatic growth and develop-
    ment. Cancer Surv 11:123–146
Streuli CH, Schmidhauser C, Kobrin M, Bissell MJ, Derynck R (1993) Extrac-
    ellular matrix regulates expression of TGF-β1 gene. J Cell Biol 120:253–
    260
Sylvester SR, Morales C, Oko R, Griswold MD (1991) Localization of sul-
    fated glycoprotein-2 (clusterin) on spermatozoa and in the reproductive
    tract of the male rat. Biol Reprod 45:195–207
Tenniswood M, Guenette RS, Lakins JL, Mooibroek M, Wong P, Welsh J
    (1992) Active cell death in hormone dependent tissues. Cancer Metastasis
    Rev 11:197–220
Tenniswood M, Taillefer D, Lakins J, Guenette RS, Mooibroek M, Daehlin L,
    Welsh J (1994) Control of gene expression during apoptosis in hormone
    dependent tissues. In: Tomei LD, Cope FO (eds) Apoptosis II. The molecu-

lar basis of apoptosis in disease. Cold Spring Harbor Laboratory Press, Cold Spring Harbor, pp 283–311

Tenniswood MP (1986) Role of epithelial-stromal interactions in the control of gene expression in the prostate: an hypothesis. Prostate 9:375–385

Tenniswood MP, Montpetit ML, Leger JG, Wong P, Pineault JM, Rouleau M (1990) Epithelial-stromal interactions and cell death in the prostate. In: Farnsworth WE, Ablin RJ (eds) The prostate as an endocrine gland. CRC Press, Boca Raton, pp 187–207

Thompson TC (1990) Growth factors and oncogenes in prostate cancer. Cancer Cells 2:345–354

Thompson TC, Cunha GR, Shannon JM, Chung LWK (1986) Androgen-induced biochemical responses in epithelium lacking androgen receptors: characterization of androgen receptors in the mesenchymal derivative of urogenital sinus. J Steroid Biochem 25:627–634

Vollmer G, Michna H, Ebert K, Knuppen R (1994) Androgen ablation induces tenascin expression in the rat prostate. Prostate 25:81–90

Walker NI, Harmon BV, Gobë GC, Kerr JF (1988) Patterns of cell death. Methods Achiev Exp Pathol 13:18–54

Wilding G (1991) Response of prostate cancer cells to peptide growth factors: transforming growth factor-beta. Cancer Surv 11:147–163

Wilson MJ, Ditmanson JV, Sinha AA, Estensen RD (1994) Plasminogen activator activities in the ventral and dorsolateral prostatic lobes of aging Fischer 344 rats. Prostate 16:147–161

Wilson MR, Easterbrook-Smith SB, Lakins J, Tenniswood M. (1995) Mechanism of induction and function of clusterin at sites of cell death. In: Harmony J (ed) Clusterin: function in vertebrate organ development, function and adaption. Landes, Austin (in press)

Wong P, MacDonald IM, Sood R, Smith C, Pilon R, Tenniswood MP (1993a) Identification and partial characterization of a candidate gene for X-linked retinopathies using a lateral approach. Genomics 15:467–471

Wong P, Pineault JM, Lakins J, Taillefer D, Léger JG, Wang C, Tenniswood MP (1993b) Genomic organization and expression of the rat TRPM-2 (clusterin) gene, a gene implicated in apoptosis. J Biol Chem 268:5021–5031

Wong P, Taillefer D, Lakins J, Pineault J, Chader G, Tenniswood M (1994) Molecular characterization of human TRPM-2/clusterin, a gene associated with sperm maturation, apoptosis and neurodegeneration. Eur J Biochem 221:917–925

Woynarowska B, Wikiel H, Bernacki RJ (1989) Human ovarian carcinoma beta-N-acetylglucosaminidase isoenzymes and their role in extracellular matrix degradation. Cancer Res 49:5598–5604

Wyllie AH (1987) Apoptosis: cell death in tissue regulation. J Pathol 153:313–316

Yan G, Fukabori Y, Nikolaropoulos S, Wang F, McKeehan WL (1992) Heparin-binding keratinocyte growth factor is a candidate stromal to epithelial cell andromedin. Mol Endocrinol 6:2123–2128

Ying TS, Sarma DSR, Farber E (1980) The sequential analysis of liver cell necrosis. Am J Pathol 99:159–174

Zakeri Z, Curto M, Hoover DM, Wightman K, Engelhardt J, Smith FF, Kierszenbaum AL, Gleeson TG, Tenniswood M (1992) Developmental expression of the S35-S45/SGP-2/TRPM-2 gene in rat testis and epididymis. Mol Reprod Dev 33:373–384

Zakeri Z, Bursch W, Tenniswood M, Lockshin R (1995) Cell death: programmed, apoptosis, necrosis, or other? Cell Death Differ (in press)

Zhu X, Ling N, Shimasaki S (1993) Cloning of the rat insulin-like growth factor binding protein-5 gene and DNA sequence analysis of its promoter region. Biochem Biophys Res Commun 190:1045–1052

Subject Index

Ernst Schering Research Foundation Workshop

Editors: Günter Stock
        Ursula-F. Habenicht